COMMUNICATION, DEVELOPMENT AND DEMOCRACY

Mapping a Discourse

THE HAMPTON PRESS COMMUNICATION SERIES
International Communication
Richard C. Vincent, supervisory editor

Goodbye Gweilo: Public Opinion and the 1997 Problem in Hong Kong
L. Erwin Atwood and Ann Marie Major

Democratizing Communication: Comparative Perspectives on Information and Power
Mashoed Bailie and Dwayne Winseck (eds.)

U.S. Glasnost: Missing Political Themes in U.S. Media Discourse
Johan Galtung and Richard C. Vincent

Communication, Development and Democracy: Mapping a Discourse
Sujatha Sosale

Global Productions: Labor in the Making of the "Information Society"
Gerald Sussman and John A. Lent (eds.)

Towards Equity in Global Communication: MacBride Report Update
Richard C. Vincent, Kaarle Nordenstreng, and Michael Traber (eds.)

Reconvergence: A Political Economy of Telecommunications in Canada
Dwayne Winseck

Political Economy of Media and Culture in Singapore
Kokkeong Wong

forthcoming

Down There and Up Here: Orientalism and Othering in Feature Stories
Elizabeth Eide

COMMUNICATION, DEVELOPMENT AND DEMOCRACY

MAPPING A DISCOURSE

Sujatha Sosale
University of Iowa

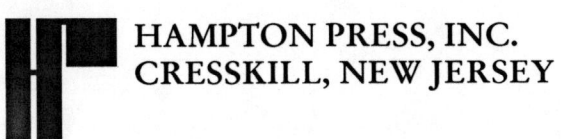
HAMPTON PRESS, INC.
CRESSKILL, NEW JERSEY

Copyright © 2008 by Hampton Press, Inc.

All rights reserved. No part of this publication may be reproduced, stored in a retrieval system, or transmitted in any form or by any means, electronic, mechanical, photocopying, microfilming, recording, or otherwise, without permission of David W. Shave, MD.

Printed in the United States of America

Library of Congress Cataloging-in-Publication Data

Sosale, Sujatha.
 Communication, development and democracy : mapping a discourse / Sujatha Sosale
   p. cm. (The Hampton Press communication series)
 Includes bibliographic references and index.
 ISBN 978-1-57273-802-7 (hbk.) -- ISBN 978-1-57273-803-4 (pbk.)
 1. Communication--Social aspects. 2. Mass media--Social aspects. 3. Communication in economic development. 4. Communication and technology. 5. Communication policy. 6. Communication, International. I. Title

HM1206.S657 2008
302.209172'4--dc22

2007041058

Hampton Press, Inc.
23 Broadway
Cresskill, NJ 07626

CONTENTS

Preface	vii
Acknowledgments	xv

1 Introduction	1
Development, Democracy, and the New World Information and Communication Order	3
Mapping a Discourse of Communication, Development, and Social Change	9
Policy as Discursive Space: Reading Development in the NWICO Debates and After	14
Chapter Summaries	17
2 Projects and Proposals for Communication and Development	21
Enumeration: Taking Stock and Spelling Out Strategies	25
Rituals and Archives	26
Infrastructures and Institutions	27
Organization and Management of Communication for Development	33
"Macrospeak": The Big Picture	37
Generalized Rhetoric and Inclusive Address	39
Cooptation of Participatory and Expressive Cultural Practices	42
Conclusion	45
3 Third World Accountability for Development Through Communication	47
Surveillance: Overseeing Communications, Overseeing Development	49
Satellite Communication and Transparent Economies	51
Communications Research as Lessons for Development	55
Accountability in Project Evaluations and Progress Reports	57

Invisibility: Hidden Overseers	60
Guiding Development Communication Projects	62\
Parent Nations as Adult Communicators	63
Communications and the Mystique of the Deferred Society	65
Conclusion	67
4 Negotiated Readings of "World Order" and "Communication"	**69**
Restructuration: Reconfiguring the Center and Periphery	73
Constructing Another Center	76
Restructuration Through Journalistic Interventions	79
Retheorizing: "Another" Understanding of Communication	81
Expanding "Public Opinion" and "Information"	83
Returning "Sociality" to Communication	85
"Alternative Communications"	89
Conclusion	93
5 From Communication and Development to Postdevelopment and Globalization	**95**
From Communication and Development to Culture and Globalization	96
Postdevelopment? Communities, Social Movements, and Local Media	103
Example 1	104
Example 2	105
Communication as Human Praxis in Global Connections	107
Development, Postdevelopment, and Globalization	109
Conclusion	112
6 Conclusion **113**	
New Media, New Developments, and Old Questions	115
References	121
Author Index	131
Subject Index	135

LIST OF TABLES

2.1.	Enumeration	25
2.2.	"Macrospeak"	38
3.1	Surveillance	50
3.2	Invisibility	62
4.1	Re-structuration	75
4.2	Re-theorizing	82
5.1	Communication and Development in the International and Global Eras	111

PREFACE

This book is about two very familiar terrains in the area of international/global communication that are so closely entwined that they seem to have emerged almost simultaneously as major conceptual bases for thinking about the field—the New World Information Communication Order (NWICO) debates for communicative democracy, and communication for development. I have treated international/global in the previous sentence as replaceable alternatives because the communication concerns expressed by the international community even prior to the NWICO debates actually pertained to globalization—a phenomenon that is now being aggressively pursued in multiple disciplines with varying degrees of reference to communication and information technology. There was a prescience in those debates that serves us well at this time.

At first flush, the NWICO debates for communicative democracy and development seem well-worn policy, activist, professional, and intellectual paths to tread. Democracy and development are frequently seen as two distinct roads, occasionally intersecting, but still parallel, and generally leading to the same destination—a world that is inscribed in the collective imaginary as modern and developed. In this project, I argue that the two paths are in fact intertwined. Given the volume of research available in both these areas, the project then becomes almost limitless in scope because the idea of development can be traced back at least to the era of colonialism if not earlier, and the idea of democracy even farther back in history. Therefore, I decided to study what development means and where it stands in communication and its concerns with democracy, using the NWICO debates as an entry point and (an albeit recent) historical marker to limit the scope of the study. Beyond the identified approximately 12 particularly intense years for this debate (1974-1985), I attempt to understand the continuities and changes in what is now often referred to as the postdevelopment era, and the implications of the latest developments in media for communication and social change at a time when we continue to have a deeply divided world based on multiple inequities.

Mapping a discourse is an intellectual approach to the practical problems of communication, development, and social change. But discourse is an interesting concept—a totality that encompasses theory and practice and, as I argue later in the book, praxis. It is also a personal journey, encoding the researcher's own orientations and experiences within its confines. Interpretations of texts in this study are post hoc, at times demonstrating what might seem a too-neat reasoning that only the luxury of hindsight offers. Also, to an extent, such interpretive exercises are unique to the interpreter who originates them and always at best a partial picture. This implies an outcome that may not be in accordance with the pictures and visions of other members of the development constituency at large. But no picture is complete; each view adds a piece to the puzzle, and I hope this work is treated as part of a larger dialogue about the puzzle and practices of development communication. My origins and formative years in a developing country, India, and my experiential locations between tradition and modernity have contributed to a certain sensitivity to notions of development and the term's hierarchical connotations. Negotiations between the two on a daily basis, often in minute but powerfully influential ways, were followed by geographical distance (to academia in the United States) that allowed for putting previous experiences into perspective. The journey has been an experience that has provided answers to old questions and raised new ones. Thus, in some senses, *Communication, Development and Democracy* is more than a scholarly work. On some level, it is also a personal exploration of a conundrum that will, hopefully, resonate with the experiences of many people active in this field, both from so-called developing and developed regions.

OVERVIEW OF THE PROJECT

The centrality of the idea of development persists in popular and élite perceptions worldwide. Equally central is the concept of democracy to media and communication studies. In scholarship, democracy is debated in more philosophical and theoretical terms; in what is generally considered a more practical realm, the struggle for democracy finds expression through policy and community action. When these communication issues are extended to the international/global arena, often the concepts of development and democracy become historically contingent social states of being around which much collective anxiety, political debate, and monetary assistance center. The emergence of a discourse of communication, development, and social change from debates about the recognition of democracy as a basic need forms the object of this study. The aim of the project is to examine how development plays a central role in deliberations about democracy in world media communications.

If one were to plot graphically the recent trajectory of the related struggles for global democracy and development, it would reflect heightened activity during the debates for a New World Information and Communication Order (NWICO) in the decade of 1976-1985 involving the various nation-states, recede thereafter, and resume the upward curve now with the World Summit on the Information Society (the WSIS) sponsored by the United Nations and organized by the International Telecommunication Union, with the private sector and civil society constituencies also playing a role. This project follows this trajectory, with particular attention to the decade of 1976-1985 when NWICO debates were at their most intense. It then provides a conceptual location for communication and development in the global era and attempts to reinterpret development and democracy with particular reference to the perils and promises of new media in the information society. This said, the project by no means attempts a complete history of communication and development, but deals with general themes of development embedded in the bid for global communicative democracy in the last thirty years or so. In this context, I examine the roles of technology, economics, and culture as constitutive elements of a discourse of communication and development that has emerged at least in part over the last 35 years.

Studies in international communication frequently signal the NWICO debates as a point of departure for a formal bid for the establishment of global democracy. It is a particularly illustrative period in recent history where developing regions were preoccupied with serious efforts to establish a more fair distribution of communicative power in the international information and communication sphere. The tug-of-war settled around two positions. On the one hand, the developing regions proposed state and other interventions to ensure a balanced flow of information in the global arena that was critical to ensuring a basic need—democracy. On the other hand, core developed nations categorically opposed state intervention and management of the global symbolic arena on grounds that the market, and not the state, arbitrated the most realistic and fair democracy.

Against this backdrop that emerged in the 1970s is the larger struggle for instituting social change worldwide to improve conditions of developing societies, and it is here that this project enters the picture. This struggle was characterized in most visible and voluble terms by the management of and intervention in the "unruly terrain" (Crush, 1995), as the world is perceived. The general tenor of expectations on the side of advanced nations was that democracy would follow development, whereas developing regions saw global communicative democracy relatively independent of development. I demonstrate that a discourse of communication and development can be read in this tangle of expectations, at times imagined antecedents, and many real consequences that mark international/global communication.

Using the NWICO debates as an entry point, this book seeks to examine communication, development, and social change as they texturized and layered the democracy debates in the NWICO. It then takes the discussion to an understanding of the extent and nature of change in the discourse of communication and development in the global era, and finally the role of new communication technologies today in the making of global democracy and development. Mapping a discourse requires accounting for discursive constructions as well as contestations. Mindful of this dialogic relationship, I also account for resistance to and negotiation with the dominant discourse of communication and development by a variety of actors at different levels (intergovernmental organizations such as the UNESCO, regional-level consortia such as the Nonaligned Movement, national projects, development corporations, and grassroots communities). I derive the method and mode of analysis for this project from two theoretical streams that have begun to acquire increasing prominence in international communication—critical cultural studies, which is concerned with social power formations, and poststructuralism which, in a larger sense, accounts for the discursive formations that articulate social phenomena. I plot a series of tropes (following White, 1978, who noted that tropes are the "soul of discourse") that emerged around the NWICO debates. These tropes detail the mechanisms constituting the discourse. I then examine historically and conceptually the shift from development in the international era to postdevelopment in the global era. I end the volume with a look at new media, the Internet in particular, that appear to emerge now as representative of new networks of development and democracy.

The texts examined for this project vary considerably in range. From selected policy documents, research studies, and minutes of meetings generated in the UN and UNESCO meetings around the NWICO, to meta-theorizing from the literature and selected studies, to selected Web sites, the analysis ranges over a series of sites where the discourse of communication and development has unfolded over the past 35 years. Mapping a discourse by identifying its outer bounds is not possible because the material generated at this time in official, scholarly, professional, and other circles was prodigious (the UNESCO publications index count alone amounted to more than 2,000) that realistically pursuing a more modest goal of drawing out some of the discursive contours using selected works, both better known and less known, would be more productive. Some of the selected texts pose a paradox to understanding the discourse of communication and development. Indeed UNESCO, an important discursive site for this study, poses an analytical dilemma. It possesses a record of initiating, supporting, and encouraging empowering projects in developing nations. Experts who have worked in UNESCO projects are also committed activists, in a sense. They have brought and continue to bring their experiences and expertise to the table, wrestling with seemingly insoluble communication-, culture-, and development-oriented problems. Even so, there are boundaries and constraints (tangible and real) within which deliberation and action must occur in this site. Such boundaries emerge from a combination of funding sources, the composition of member countries, and clashing agendas characteristic of an institution that is as much a bureaucratic structure as it is an organization given to working for peace and social change. The discourse emerging from such a site inevitably contains both ideologically dominant as well as negotiatory positions on what the idea of communication, development, and democracy means.

Particularly during the time period selected for the study, under the leadership of Amatar M'Bow, many UNESCO-commissioned studies generated perspicacious and insightful works that straddled what I read as both a traditional developmentalist discourse as well as radically innovative paths to new ways of thinking about and doing development. Other works outside the UNESCO also negotiate with this discourse to offer new alternatives. Many of them remain a significant contribution to the corpus of work on communication and development for multiple reasons. But at the present time, we cannot overlook the fact that some of these texts at times carried strains of support for a discursive power they were also fighting. Reading these texts posed a challenge; they appear in all areas—dominant, contestatory, and negotiatory. For analytical purposes, the dialogic relationships here have been artificially separated to read what I have identified as tropes. This separation allows for a closer examination of certain threads in the fabric of the discourse. This range of texts and the multiple modes of analysis (both conceptual and empirical) enable me to trace processes in the formation of a discourse of communication and development in relation to global democracy that, although contested, managed to retain the power to normalize a particular type of ideal developed democratic society.

The book consists of six chapters, beginning with an introduction to the history, positions, theories, and method pertinent to the project. In the second, third, and fourth chapters, I closely examine the discourse of development in the NWICO debates through two sets of tropes for each chapter—enumeration and macrospeak, surveillance and invisibility, and restructuration and retheorizing. These tropes demonstrate some dominant and contestatory positions on communication, development, and social change. The fifth chapter addresses the move from an international development era to a global postdevelopment era from within a historical framework; it examines implications for communication and development in a globalizing world. Finally, in the last chapter, I examine briefly the debates and activities surrounding the World Summit on the Information Society and its efforts to address a new development problem—the digital divide. This chapter situates the discourse of communication and development in relation to a version of cyberdemocracy today.

A note on terminology used in the project—the term communication takes a broad perspective on communication and media. The term information, generally associated with mathematical theories or decision making in an environment of uncertainty, is treated differently here. It has been extended to the symbolic domain that is associated more with communication (in keeping with works in journalism, journalism training, measures of communication capacities, etc.). At times in the study, information and communication are used interchangeably. Perhaps the most difficult term is Third World (and its corollaries of First and Second Worlds, and, in one instance in chap. 5, a Fourth World). Partly, the divisions hold because the period selected for the study predates the global sea changes in 1989 and after. For development anthropologists such as Escobar, the term is "a regime of geopolitical representation" that continues to conceptualize the First and Third Worlds as North and South, an economic divide that was a sensitive topic in the NWICO debates (Escobar, 1995, pp. 109-110). The term, when used in this study, is also mindful of the present hierarchical chasms between the two worlds in various domains.

LOCATING THE PROJECT IN THE (POST)DEVELOPMENT DIALOGUE

This project deals with development as unfolding against the backdrop of notions of global democracy. It engages with ongoing debates in the critical development community from a communication perspective that is located at the heart of both concepts—development and democracy. It differs from other works on development and democracy in specific ways. Rather than focusing on specific practices of development communication, such as the diffusion of innovations or development journalism (although they are integrated into the discussions as constitutive of the larger discourse), or analyzing in depth democracy in relation to global information and communication flows (excellent [and extensive] work is available on both these foci as well), this book attempts to build on the fragmented but important ideas of development and democracy debated at this time in the UNESCO and other related venues. It brings to the surface the processes and mechanisms by which development underwrote, and continues to underwrite, the quest for global democracy. Additionally, this project lays a foundation for future work in development and democracy, particularly in the wake of the WSIS, and the renewed interest and urgency in the questions of development and democracy that have appeared in the UNESCO, now with a new component that was noticeably absent in the earlier NWICO debates—civil society constituencies.

Contemporary scholarship in communication and development increasingly favors critical perspectives, and this project can be located in this neighborhood. At present, urgent attempts are being made to revisit the global democracy question in light of the latest technological developments in the information age. Existing literature mostly tends to pursue one or the other question—global development or global democracy. Questions of democracy, at the heart of public expression in mediated symbolic space, have had a history of being embedded in certain notions of development. Therefore, it is vital to investigate a process that organizes economies and social imaginaries as developed/developing, media-rich and media-poor, or as Schiller (1999) has accurately described the digital divide, "informationally-privileged and informationally impoverished." Ultimately, at the level of the everyday in various communities across the world, these questions directly address the battles between power and empowerment and limits and possibilities for global democratic expression.

ACKNOWLEDGMENTS

This book has taken shape over the course of the past few years. Singling out people to whom and opportunities for which I am grateful will inevitably lead to omitting some (I suspect key) people and sources. I have tried to list comprehensive groups of people who have influenced the project and have been supportive of it. For others who may not figure here—my apologies—and my gratitude is no less. First, I am deeply indebted to the many scholars, policymakers, and activists in media policy, media studies, development studies, history, theory, philosophy, and other related disciplines who have given researchers in many fields inspiring and productive insights through their interdisciplinary works—a vast and rich reservoir of knowledge and wisdom out of which a small fraction is referenced in this project. Equally, my gratitude for the many works in this study emerging from so-called Third World experiences, some of which I have, in places, contested. These works are no less valuable for their perceptive and astute understanding of non-western media situations, institutions, discourses, achievements, and roles, and have taught me much about important media forms and development efforts that would otherwise be easy to miss. I also hope that I have done justice to their contributions to questioning dominant ideologies of development. I have engaged with several of these works and with the literature so intensely that at times it is difficult to say where their ideas end and my own begin.

I would like to acknowledge the following people for their direct contributions to the subject matter—to Dr. Richard Vincent, editor of the series to which this book is an addition, for his expertise and experience with the subject matter as well as the publishing process—he flagged potential substantive and other pitfalls; to my professors in the University of Minnesota for their feedback on very early forms of this project. Papers from this project were presented at various conventions, and to the reviewers and audiences there, my sincere thanks. To my publisher Barbara Bernstein, much gratitude is due for her patience and guidance through the publishing process. I am deeply appreciative of the support and encouragement of my colleagues at Georgia State University and The University of Iowa. I am grateful for the research funding at Georgia State University that helped me work on the project over the summer. A subvention grant from the Office of the Vice President for Research, The University of Iowa, contributed toward indexing costs. I am grateful for this timely help.

I would like to thank the following publishers for copyright permission—material from these sources has been excerpted and modified in certain chapters, to suit the flow of the arguments in the book. The order presented here reflects the order in which these works have been integrated in the book.

Sosale, Sujatha. (2004). Toward a critical genealogy of communication, development, and social change. In Mehdi Semati (Ed.), *New frontiers in international communication theory* (pp. 33-53). Lanham, MD: Rowman and Littlefield.

Sosale, Sujatha. (2003). Envisioning a New World Order through journalism: Lessons from recent history. *Journalism, 4*(3), 377-392.

Sosale, Sujatha. (2003). The panoptic view: A discourse approach to communication and development. In Jan Servaes (Ed.), Approaches to development: Studies on communication for development. Paris: UNESCO Publishing.

Sosale, Sujatha. (2002). Communication and development in the international and global "eras:" Understanding continuities and changes. *Journal of International Communication, 8*(2), 8-25.

Last, but not least, my acknowledgments go to my wonderfully supportive family. This book is dedicated to them. They encouraged, chided, and celebrated at the various stages of development of this project. To my mother Rajalakshmi and my father, the late Srinivasagopalan, my sincere thanks for their patience with the difficulties that accompany a long-term project and faith that it would be completed. My mother's help and support at the final stages of completing the manuscript were most timely. My sincere thanks to my sister Shobhana and my brother Dinakar, for their unflagging support, encouragement, and other kinds of help too numerous to list here. They shared in my anxieties and triumphs. This project is so much the better for them.

INTRODUCTION

There is a tension in academic and policymaking arenas now between what might be identified as two broad constituencies. The one claims that the world has passed into an era of globalization, rendering questions about development (at least in their familiar forms) irrelevant. The other continues to struggle with issues of development, tracking their consistencies as well as changing forms and problems in the current globalizing world. Both constituencies are intensely aware of the composition of the global society—predominantly the First, Third, and Fourth Worlds, with the Second World considerably diminished in discursive prominence after the political sea changes of 1989. The first constituency explains the present global social composition through a combination of global capitalism, information technologies, migratory labor, significantly increased international tourism, intensified migration by refugees, and (re)migration to identified ethnic homelands. In contrast, the second constituency attempts to understand current changes, shifts, and trends through the lens of development/underdevelopment, sustaining the discourses of technology and economics, cornerstones of the larger development discourse. As part of attempts to understand changes and shifts, the second constituency also grapples with the consequences of globalization and underdevelopment at the level of everyday lived experiences. These epistemologies are responsive to the current global era, carried over from the post-Second World War development/ underdevelopment divide in the second half of the 20th century to the new millennium. The two constituencies do not always validate or dismiss each other; they exist as two tracks, at times meeting at certain intersections (as is evident in Tehranian's [1999] work). Both constituencies produce competing and persuasive versions of recent transnational history and the current global reality.

This study claims membership in the second constituency, and it engages with questions of development and social change, specifically from the disciplinary perspective of media and communication studies. It also accounts for the shift from an international, development-oriented world to a world oriented to global capitalism as a historical phenomenon from its position within the community of concerned scholars (and now as is increasingly publicly acknowledged—activists and practitioners) in the development constituency. The aims of this book are threefold: (a) to account for the processes by which the idea of communication, development, and social change acquired certain meanings at the expense of possible others at a specific time in international communication history; (b) to understand the implications for communication and development in the shift from the international to the global eras; and (c) to examine approaches to the role of communication in social change in what is now being construed as a postdevelopment age, leading to suggestions for new directions for inquiry in the area of communication and global social change.

Implications for democracy in each of these three goals constitute a running thread and appear either to inform from the background or are periodically foregrounded in the construction of the discourse of communication, development, and social change. Although numerous other (many excellent) volumes in the past, both recent and not so recent, have provided what might read as similar accounts, this project parts company there and begins with the premise that to study communication is to study democracy, and to study communication in the international/global arena is to address the global hierarchy historically wrought by development and its relation to democracy. At the heart of the mediated communicative act is the notion of democracy, largely referring to equal opportunity for public expression, access to public expression, and fair representations of social groups in the symbolic public sphere. In the international arena, with the creation of the United Nations, the motto that all nations are equal, and that all member nations are "expected to observe democratic sentiments" (Berger, 2001, p. 217) informed the debates over global distributive justice of communication resources, opportunities, and representations. Historically, opportunity structures for communication, access to communications media, and the power to represent peoples to national and global publics have been directly tied to the level of development of a given nation at a given point in time. It is conventional wisdom now that wealthier nations possess a disproportionately large share of the means to communicate and represent. The collective finger, therefore, has been pointed to development as enabling democracy.

Against this backdrop, I first approach the concept of development through the recent New World Information and Communication Order (NWICO) debates in the UNESCO to map the discourse of communication and development that underwrote efforts to establish a communication policy for fairness in communication flow and representations of nations and peoples in the global communication circuit. I then follow the discourse of communication and development into the globalization process and examine the implications of this process for development and democracy in the present global era. I develop this line of thought by looking generally at the informatization of society as conceptualized broadly by the discourses surrounding the World Summit on the Information Society (WSIS) and, more specifically, at initiatives both at the level of transnational e-commerce and within new media technologies that are engaged in what has been called globalization from below. The WSIS is a significant event in that the old players in the NWICO debates, UNESCO and the International Telecommunication Union (ITU), debate the global information society, this time with a difference—unlike in the NWICO debates, a large representation from civil society has actively participated in the WSIS.[1] The discussions on the global digital divide and its implications for developing regions featured prominently on the agenda of the WSIS summits. With the arguments supporting the democratic potential of new media in the digital age informing the summit, and the old players from the NWICO debates constituting major participants, there is a sense of a *déjà vu* that makes questions of development and democracy germane to the present era.

DEVELOPMENT, DEMOCRACY, AND THE NEW WORLD INFORMATION AND COMMUNICATION ORDER

Following World War II, recognition of the centrality of modern communication in the global arena arose from three factors: (a) a need to *modernize* ex-colonies, (b) *preserve* cultural sovereignty and national identity of these new nations, and (c) enable these nations to *compete* efficiently in the new world order marked by world markets, rapid information and monetary transfers across borders, and the growth and use of advanced communication technologies to enable such transfers at a hitherto unprecedented scale

---

[1] The consortium, Communication Rights for the Information Society (CRIS), has raised doubts about ITU's plans to allow full participation of civil society groups, and sees this restriction, ultimately, as a way to confine the WSIS debates between governments and corporations, a state of affairs that prevailed during the NWICO debates.

and velocity. Thus, communication was seen as contributing to and resulting from a more global idea of development, and it was expected that the two together—communication and development— would eventually ensure a democratic (construed simultaneously as free and balanced) flow of information and communication among all nations. This idea featured prominently in the UNESCO intergovernmental deliberations over a new international information policy.

The overt and explicit content of the debates had to do with democratic communication. The imbalances in news and information flow, communication opportunity structures such as access, and the misrepresentations of certain parts of the world in the media of economically advanced countries were considered to impede a fair distribution of communicative power in the global arena. The NWICO debates attempted to address imbalances in flow and misrepresentations through pragmatic solutions. In the conflict that ensued, proponents of the policy suggested the monitoring of journalists (a major issue of contention in the debates) as one of several ways to control imbalances, whereas the opponents to the NWICO advocated the greater freedom of information flow, access, and representation to ensure, in their interpretation, true global democratic communications.

The issue of communication and development constituted a considerable part of the debates for a new information order. The role for communication and development was articulated as follows in the MacBride Commission's proposal of a NWICO. This quotation serves to highlight the place the concept of development occupied in the deliberations about instituting an information policy intended to facilitate global democracy in mediated communication:

> Development strategies should incorporate communication policies as an integral part in the diagnosis of needs and in the design and implementation of selected priorities. In this respect communication should be considered a major development resource. (UNESCO 1980, p. 28)

In 1976, at the prompting of the nonaligned countries,[2] a resolution to develop a new international information and communication order was passed in the UNESCO (Gerbner, Mowlana, & Nordenstreng, 1993). The investigations and debates around the NWICO culminated in the MacBride Report in 1980. The report contained a draft for a NWICO designed to bring communicative balance in the international arena. The NWICO proposal exacerbated existing controversies in the UNESCO until, in 1984, the United

---

[2]The nonalignment movement included a consortium of countries, mostly from the developing regions and continents, that consciously positioned themselves as a third entity, distinct from the bipolar cold war world and its constituent powers. A more detailed treatment of the NAM's role in the NWICO is available in chapter 4.

States withdrew its membership from the organization, followed by Britain in 1985, thereby reducing considerably the prominence that NWICO had acquired on the UNESCO agenda. Within the longer history of communication and development, the decade of 1976–1985 serves as a point of departure for this project—a critical point in time when the discourse of communication and development unfolded in relation to the quest for an equitable distribution of communicative power among nations.

The demand for a new international information order closely followed the demand for a new international economic order because the existing economic order obstructed the achievement of significant social change in the Third World. The new economic order was differentiated *qualitatively* from the existing economic order (Addo, 1984)—it was envisaged as being characterized by cooperation (rather than competition), negotiation (rather than domination and subordination), and an "interdependent and equitable [international] structure."[3] Communication was considered to be integral to effecting this change, and soon a demand for a new information order took shape. However, quantitative concerns over imbalances in information flow tended to override the attention to actual information content in the demands for a NWICO. A former director of the Inter Press Service pointed out that rather than directly addressing silences and absences in the existing flow, many participants of the NWICO debates attempted to adjust and "correct the imbalances" (Savio, 1982).

This is in no small way due to the pressures faced by the authors of the document consolidating the position of the proponents of the NWICO, the MacBride Report, to incorporate opponents' stands. But responses to such pressures softened the edge the debate had acquired in 1974–1980. Some participants in the debates expressed concern over the investment in communication infrastructures and technology at the expense of communication content. In the discourse of modernization, technology produces and is at the same time a product of development. Thus, development in relation to communication, and how communication could help achieve development (communication *and* development, communication *for* development, development communication, and, therefore, the need for communication development to correct imbalances in international information flow and preserve national cultures and sovereignty), constituted a considerable part of the NWICO debates. Terms such as *social progress, cultural development,* (typically referring to modernization), and *political development* (creation of a democratic climate conducive to the formation of effective public opinion) were frequently positioned as prerequisites for a balanced flow of communication. Toward the end of this decade, NIWCO opponents held firm that technologization (modernization) would eventually solve the problem of

---

[3]For a detailed and politicized analysis of communication technology and information flows, and their impact on developing countries, see Hamelink (1984).

imbalances in global information flow (the problem of democracy). Proponents' interpretations of the use of technology were tied to the idea of democracy, but autonomy in choice of technology use, access to means of communication, and moderation of representational practices in the media were seen as achieving democracy because they would simultaneously manifest vernacular developments.

Broadly, the polemics surrounding the possibility of instituting a NWICO policy include the following arguments: (a) a *laissez-faire* international media market to ensure true democracy, helped generously by free export and use of communication technologies; this, in turn, would promote a free international marketplace of ideas; *versus* (b) attempts to protect newly formed Third World nations and their indigenous cultures from perceived cultural invasions caused by the new global technologies and the powers behind these technologies. An examination of the MacBride Report shows approximately three main areas of concern expressed by developing countries; (a) democratization of information and the dependence/interdependence rhetoric, (b) professionalization of media personnel, and (c) the question of development in relation to communicative democracy. The need for a new information policy was usually attributed to imbalances in news flow, distorted representations of several parts of the world by the mainstream media of the developed nations, the influx of foreign media programs in Third World countries, and the imbalances in resource distributions, particularly in the case of electronic and satellite communications. The MacBride Report integrates communication and development with these concerns under sections such as "Expanding Infrastructures" (communication technology), "Integration" (the traditional and the modern, and combinations of new technologies), "Disparities" (differences due to disparities in economic development), and "Material Resources" (infrastructure and technology). Also included in the MacBride Report was the formal proposal for the new policy presented under the section "Communication Tomorrow."

Subsections in the policy proposal addressed communication and development for self-reliance, basic needs, the importance of "integrating communication and development," "facing the technological challenge," and social progress. The bulk of the literature on the NWICO tends to focus primarily on autonomy and interdependence in relation to information flow; the flow and cultural imperialism concerns have constituted a more vocal and politicized part of the policy debate. When development was discussed within the policy context, it was to highlight its role in the establishment of a new order. In the debates involving technology, economics, infrastructure, and media, participants also addressed the use of traditional media for political, economic, technological, social, and cultural development. Proponents of the NWICO saw a new order as more than achieving equitable distribution of communication power; they saw the policy as enabling development of newly independent nations and, accordingly, built in a demand for recognition and support of this goal in the policy proposal. Thus, development was imbricated in the information policy debate as contributing to a new information order and also as a state of affairs that would continue to derive from that order.

The prolonged history of political and economic subjugation of many newly independent nations among the NWICO advocates led them to internalize and valorize West-derived themes about the understanding and roles of information and communication in the development project. But the discursive control over terms such as *development*, *equality*, or *freedom* did not prevent constructions of alternate meanings for these terms or pursuit of alternate development practices. The NWICO debates were, for the main part, intergovernmental and were conducted in diplomatic circles, with the UNESCO serving most frequently as their institutional site. Radical opposition to the world powers was unusually loud during the 1970s. The playing out of the discourse can be charted as follows. At its most obvious, this control was exercised by the economically powerful countries in their implicit threat to withdraw funds, technology transfer, and aid for professional training. More subtly, a history of establishing Western societies as exemplars of progress, development, and the ideal society contributed considerably to this discursive control. The movement toward new ways of interpreting development in the context of the realities of the formerly colonized countries warred with their acceptance and promotion of the dominant ideology of development.

The discourse of development continues to inform institutions, practices, and expectations (Escobar, 1992). Asymmetries in global information and communication also persist (Gerbner, Mowlana, & Nordenstreng, 1993; Shah, 1996). A sizeable community of communication scholars, activists, and practitioners continues to share this concern, and this project is situated and participates in this space of shared concern. For example, some of the arguments for free trade appear in the World Trade Organization (WTO) conferences, where a powerful lobby demanding inclusion of services under trade and intellectual property rights has overlooked the unique properties of "culture industries" (Braman 1990). As Braman also pointed out, culture industries continue to be embroiled in unequal exchanges of information, and information and communication have been increasingly commoditized to fit the changing concerns of the international economic community. The economic domain dominated the operations of communication at the macrosociological level (international, regional), and, as a response, the cultural role of communication was increasingly obscured. Nonetheless, the problems raised by the NWICO persist as a review of the UNESCO survey of the news exchange mechanisms (NEMs) of the global South in the early 1990s indicates, an activity that "keep[s] alive various issues raised during the NWICO debate" (Boyd-Barrett & Thussu, 1993, p. 186). The constraints against which these NEMs continue to operate with limited effectiveness point to the enduring problem of significant inequity in global communication exchange and its links to the discourse of development. Such inequities have survived discourses of development into the new information society (Bailie & Winseck, 1997). Some participants in the debate conceptualized the NWICO as "an idea, and a process—not an event or a program" (Gerbner, Mowlana, & Nordenstreng. 1993, p. 1). Further, they shifted the scope of the new order from an inter-national debate to its more abstract, ethical, and humanistic mission—". . . cultural emancipation, political democracy, and realization of peoples' right to self-determination" (Gerbner, Mowlana, & Nordenstreng, 1993, p. 1). This ethical mission, now at the center of contemporary globalization studies, reinforces the importance of the NWICO as a critical entry point to study the discourse of communication and development in the last three decades.

The NWICO debates were confined mainly to intergovernmental circles. On the infrequent occasions when they covered the UNESCO meetings, mainstream commercial media in many countries opposed the move toward such a policy (Mattelart, 1994). A key to understanding the construction of the discourse is located mainly in the official rhetoric recorded in UN documents and reports. Drawing partly from conceptual work on official discourse and the discourse of official ritual, and utilizing a variety of sources such as the UNESCO and other UN publications, studies conducted for the MacBride Report, selected papers from the nonaligned countries' media conference (Final Reports and Documents of the NAMEDIA Conference, 1983), and academic journals, I account for the role of official rhetoric also in constructing a discourse of communication, development, and social change.

Attempts to map a discourse of communication and development with the NWICO debates as the starting point for this project raise the following questions: What strategies are apparent in constructing the discourse? How were negotiations with the dominant discourse conducted? How did proponents of the NWICO employ the concepts in the dominant discourse to advance vernacular versions of communication and development? Were and if so how different were these meanings and interpretations from their modern (i.e., development in its late 19th and early 20th century forms) origins? The questions then carry to a continuity and change theme from the onset of present-day globalization that, arguably, commenced in the 1980s.

To attempt answers to these questions, I begin with a theoretical framework located at the intersection of critical cultural studies and poststructuralism. This framework is important in that it helps explain the establishment of modern communications and modern development as an arbitrary yardstick of the ultimate quality of life for about three fourths of the global population, and the resultant implications for considering communication as an expressive cultural practice and a democratic act—whether through traditional or vernacular modern modes and channels—rather than as a tool integrated into the disciplining power of development.

MAPPING A DISCOURSE OF COMMUNICATION, DEVELOPMENT, AND SOCIAL CHANGE

A historical dialectic of sorts has characterized the notion of communication and development for social change over the last half century or so. This notion can be conceptualized as a discourse—an entity that embodies the power to confer meanings by establishing what Hall (1997) elsewhere called as "classifications of social intelligibility." Tracing the discourse would involve paying close attention simultaneously to what Young (1981), in his interpretation of Foucault, explained as " 'critical' analysis (which examines the functions of exclusion)" and "'genealogical' analysis (which examines the formation of discourses)" (p. 49). Critical and poststructuralist approaches jointly inform for the conceptual apparatus for this study. In a critical reading of Foucault, Said (1978) emphasized the importance of and use for Foucault's notion of discourse in the international arena. But, he contended, Foucault's analyses of discourses were ethnocentric, limited to the Western and, more specifically, French-speaking world, and observed that Foucault's idea of discourse was European when indeed, by default, it should be much more. Said observed:

> ... along with the use of discipline to employ masses of detail ... discipline was also used to administer, study, reconstruct—and subsequently to occupy, rule and exploit—almost the whole of the non-European world. (p. 711)

Spivak (cited in Slater, 1992) also acknowledged the global applicability of these theories, but questioned the absence of the constitutive role of empire and the colony in the discourses emerging from these histories. Slater cited Spivak's critique of Foucault's Eurocentric or, more closely, Francocentric archaeological foci that have prevented a "reading of the broader narratives of imperialism. . . . To buy a self-contained version of the west is to ignore its production by the imperialist project" (Spivak; cited in Slater, 1992, (p. 285). Studying the discourse of communication and development calls for an engagement with an enterprise that has been frequently characterized as an extension of colonialism and as a phenomenon set squarely in the international arena. In a sense, my study is a response to this call to account for the production of the discourse of communication, development, and social change, and the dialogic nature of the discourse.

Drawing broadly from critical cultural studies and poststructuralist thought, I develop a framework to map a critical genealogy of communication and development, using the NWICO debates as an entry point into the discourse. Doty (1996) employed the term *critical genealogy* to define her project on historical representations of North–South relations, encapsulating the critical perspective associated with concepts of ideology and hegemony and the Foucaultian approach to discourse through genealogy. Specifically, the term *critical* for this project refers to problematizing centralized forms of power manifested in meanings, institutions, and practices. In using the term *genealogy*, I refer to a historical *approach* to studying communication and development as a discourse. However, this study does not claim to trace the actual longer genealogy of the idea of communication for development. The genealogy here pertains to development as it underpinned debates about democracy, especially over a decade or so, and is mapped by also examining what Escobar (1999), in explaining the relevance of Foucault for development studies, termed nondiscursive practices—such as the social, economic, and institutional. The discourse may be socially dispersed in various institutions, practices, time periods, and sentiments. Nevertheless, a "structured, relational totality" (Doty, 1996, p. 6) can be traced in this dispersal of power, and the resultant understanding and articulation of this discourse points to the definition of a specific reality. This totality, however, does not imply discursive closure; it is a particular reality that is at least temporarily fixed (and therefore lived as such) and continually contested. As Saunders (2002) has cautioned, "it is important to think power as coherent in spite of the risks of any such representation; at the same time a critical vigilance ought to be observed to avoid the trap that we've grasped a transparent reality" (p. 10). The multifarious practices of development have, over time, produced a knowledge that has come to define a truth about the status of societies and cultures. Such a framework points to both the persistence of the development enterprise as well as the equally persistent alternate modes of understanding and acting for social change.

Scholars have problematized communication and development from various perspectives such as participatory communication (Huesca, 2003; Jacobson & Kolluri, 1999), indigenous media (Michaels, 1994; Rodriguez, 2001a), humanitarian journalism and communication (Shah, 1996; Tehranian, 1999), human rights and communication (Servaes, 1998), feminist perspectives, and spiritually oriented and world religions-derived standpoints on communication and social change (Rodriguez, 2001b; Steeves, 2001; Tehranian, 1999). These recent works and their precursors signal the co-existence of alternate paradigms for communication and development.

Theoretically, the integration of postmodern and poststructuralist thought and their critique of the metanarratives of development and social change have been considered in recent works (see e.g., Shah, 1997; Tehranian, 1999, for excellent summaries; see Jacobson, 1996, for a critical analysis of these new theoretical approaches and their critique of the metanarrative of development).

However, this project differs from these works in that it contributes to three relatively unexplored areas. First, it extends the previously mentioned summary works (some more extensive than others) by examining the roles of technology, economics, and culture as constitutive elements of a discourse of communication and development. To this end, the study attempts to account for the mechanisms through which the major reference points for organizing knowledge about societies in the international arena from communication perspectives have emerged. This vein of scholarship in international communication is yet to be fully developed. Second, it accounts for the positions and perspectives articulated in parallel with and in response to the development discourse. These positions and perspectives have now come into their own as reflexive and ethical dialectic responses to mainstream notions of communication and development. I attempt to cast these new voices as a collective *other* that is integral to the notion of this discourse as a whole (Derrida, 1986). This "other" details the power/knowledge nexus that informs our understandings of a world order. Finally, this study extends the discursive domain of communication and development to the global era. I attempt to sort out major positions on communication, development, and globalization by examining both the conceptual debates as well as empirical instances of continuing efforts toward social change in the global arena and the central role of media communications in such efforts. Contemporizing the discourse allows us to grasp emerging trends, changes, continuities, and praxis that confront the needs for social change at both global and local levels.

There are two parts to the analytical framework developed for this project. The first part consists of mapping a "unity of fields" (Foucault, 1972) that characterizes a discourse. The second part involves a deconstruction of this unity of fields and, in a sense, constituting the discourse through its other. Two concepts—centering and ambiguity—are central to the task of mapping a unity of fields that constitute the discourse of communication and development as it was manifested in the NWICO debates. By centering I refer to the efforts and acts that help foreground communication and development in the collective imaginary of populations and governments. The idea of ambiguity allows us to understand the fissures and ruptures that are constantly present in the dominant discourse. By the other I mean a collection of marginalized voices, ideas, meanings, and practices that effect discursive contestation.

Establishing a particular set of meanings and practices that determine expectations and set criteria for evaluating societies demonstrates the work of power in centering meanings and naturalizing them in the social imaginary. For example, firmly established universal criteria such as production and/or use of advanced media technologies for evaluating a society as developed or otherwise articulate the formation of a center. Also, contradictory strands are woven into the construction of a center in that they share a "common locus" (Foucault, 1972, p. 152). Several elements of the major theories of development are foregrounded in the NWICO debates. The demand for such a policy emerged from the realization that information and communication, in the era of satellite and computer technologies, called for action on par with a similar demand for a New International Economic Order (NIEO). Such technological advances are typically associated with modernization theories. Analyses of communication industries, content, transfer of technology and training, and the flow of news and information revealed the need for a new kind of decolonization—the decolonization of information. This argument resonated to the implications in the dependency school. For example, in 1985, during the Information Technology and Sovereignty meeting at Yamoussoukro, the Ivory Coast, member countries resolved to pool more than 1% of their collective GNP for acquisition and management of information and information technologies for two related purposes—regional and continental integration of Africa (with the implication of delinking from dependence and moving towards self-sufficiency), and survival in the international community of the 20th century (for which modernization of information systems was mandatory). Dependency-based as well as modernization-based discursive strands engage in a dynamic interplay in the NWICO debates. The technosocial discourse of modernization constituted a large part of the debates and, to a lesser extent, the economics-based discourse of dependency. In this sense, I read from the NWICO debates the construction of a center that is communication and development.

Ambiguity is inherent in discursive formations in that it is impossible to achieve discursive closure on any ruling idea. It is the concept of ambiguity that allows for both contestation and negotiation, acts equally constitutive of a discourse as is centering. In fact we might say that the tension between centering and contestation, evident in ambiguity, is at the heart of discourse. Ambiguity compels us to consider communication and development more as a "floating signifier," to use Laclau and Mouffe's (1985) term, than as a "master signifier" (Zizek, 1993). Efforts to dislodge the master signifier create oppositional (delinking from the development project would be an example, as was the case with China or Cuba) or negotiatory meanings and actions (vernacular versions of modernity and attendant adaptations of modern communication systems and technologies that might take place on an everyday basis in various parts of the world) demonstrate the limits of the discourse while opening up possibilities for new and perhaps more effective modes of conceptualizing and communicating for social change.

Negotiation with the modernization discourse and the resulting alternate definitions of development are apparent in the cracks and fissures caused by continuous contestation and negotiation of the dominant knowledge of communication and development. Identifying these fissures leads us to inquire into the power relations involved in the discursive production of communication and development, as to the processes by which the other emerges in deconstructive readings of a dominant discourse. As Gupta (1998) reminded us, the discourse of development is "enunciated from multiple positions" (p. 43); these positions are not always obvious and are often situated in the fissures that constantly work to erode the dominant project of development.

The second part of the framework addresses the exclusionary practice of discourse. If the idea of communication and development is entrenched in the speech and action of states in developing regions and various external development agencies, it is usually at the expense of a possibility, whereby the culture and practice of communication need not be connected to development in the established sense of the term. Deconstructing key ideas in the NWICO debates helps identify and articulate an other, a collective other that encompasses more marginalized meanings and practices and negotiatory attempts of various stripes to produce alternate routes to social change.

Most of the developing nations did not see a complete disengagement with the world system as an option. Nor did they reject the idea that modern communication systems were crucial to building nations, establishing national identities, and engendering social change in desirable directions. Rather, proponents of the NWICO from member countries of the developing regions adopted these theories into their fold, inflected them with their own ideologies, and invested in them meanings that (a) had originally been attributed to terms such as *modernity* and *development*, but had subsequently been obliterated in the prolonged rush for domination; and (b) enabled a third way of seeing the concept of development—*another development* and *another communication*, or even another world system utilizing the metaphors and vocabulary of the dominant discourse (Pavlic & Hamelink, 1985)—now sometimes referred to as vernacular or peripheral modernities.

POLICY AS DISCURSIVE SPACE: READING DEVELOPMENT IN THE NWICO DEBATES AND AFTER

The genealogy of the idea of communication and development can be traced to the late 19th century when colonizers introduced media technologies for modernizing colonies first for administrative convenience and then for winning support and securing the legitimacy of their regimes among the natives (Hardt & Negri [2000] observed this of technologies in general). But this project does not account for the history of more than 100 years of the idea of communication and development. The U.S. post-World War II ideology that informed modernization and transformation of newly formed nations underpins a large part of the understanding of communication and development (and has continued in recent years, as Chakravartty [2001], Jackson & Mosco [1999], and Vincent [1997] have brought to our attention). Within this rubric, I have selected a particular discursive moment—the NWICO policy debates and the nonaligned movement's (NAM'S) member countries' contributions to the debates to examine communication and development in relation to democracy, which has competed with the concept of development to occupy center stage in international communication.

In this study, I approach policymaking efforts as a discursive space within which modes of understanding occur and meanings are constructed. With this approach, policymaking efforts acquire a historical character and develop their meaning-generating capacity through the course of debates and implementation. In the case of the NWICO policy debates, because implementation did not really happen entirely as envisaged by the proponents of such a policy, the focus is retained on the ten peak years of hope and some dialogue on achieving a fair distribution of communicative power at a macrolevel. As a form of guided social change, policy continues to play an important role in global communications, now with the discussion similar to the NWICO in many senses—the World Summit for the Information Society, beginning in 2003.

The NWICO debates reached their peak between 1976 and 1985. In 1976, the UN General Assembly adopted a resolution to promote communication foremost on its agenda and called for the UNESCO to inquire into the issue of democracy in international communication (Vincent, 1997). The inquiry by a commission appointed by the UNESCO, led by Irish ambassador to the UN, Sean MacBride, resulted in a proposal for a NWICO in the document, "Many Voices, One World: Communication and Society Today and Tomorrow. Towards a New, More Just and Efficient World Information and Communication Order," in 1980. The debates continued after the release of this report until, in 1984, the United States as a prominent member withdrew its membership from the UNESCO, and Britain followed suit in 1985. These withdrawals are generally considered to mark the closure of the debate, and the general tendency in scholarship has been to confine the NWICO debates to failure and past history. However, the need for a NWICO of sorts, albeit in changed form, continues to be recognized by others. The creation and revival of the MacBride Round Tables, 1989-1998, and the new WSIS are reminders. At this point, it would be premature to declare a desire for equitable distribution of communication in the global arena as a closed issue.

The NWICO debates were confined mainly to a restricted and elite site in the international arena. A site of this nature, through protocol alone, can establish the grounds for what is allowed to be said and in what manner it is to be said. What is noteworthy here is that within this restricted site and institutional culture, in the conflict between 1976 and 1985, opposition to the world powers' control of communication and information was expressed quite vociferously by many Third World countries and their supporters. Bourdieu's concept of the *habitus* is helpful in understanding the dynamics of discourse building in such an environment and also the play of conflict in a defined and structured environment such as the UN/UNESCO. *Habitus* refers to a system that is at once structured and structures dispositions, which tends to ensure the constancy of practices over a period of time (Bourdieu, 1994). At the same time, there is room for difference and change in anticipating the future. This combination of reproduction and change, according to Bourdieu, gives the habitus its history-making capacity. In this study, an official, intergovernmental habitus such as the UNESCO produces its own sets of practice. These practices are not unchanging; their historical legacy allows for an understanding of discourses as they are, yet changes can be read as adaptation to anticipated outcomes of the NWICO debate—be it through adoption of protective stances to maintain the existing communication order (which the United States and Britain did by withdrawing membership from the UNESCO in 1984 and 1985, respectively) or through efforts to change the status quo (a NWICO).

The texts available on this topic, from 1976 to 1985, number more than 2000 documents in the UNESCO publications indexes, more than 100 reports and studies commissioned for the International Commission for the Study of Communication Problems used to construct MacBride Commission's results and recommendations, and more than 30 entries for the NAMEDIA conference publication (report of the conference of non-aligned countries on the media—this is included here because many of the proponents of the NWICO were also member countries of the NAM and discussions overlapped during this time period. The NAM included the collective of nations adopting a third position during the cold war in a serious effort to disengage politically, economically, and historically from the cold war polarity). The NAM countries played an important role in the NWICO debates, many of them member countries of the UN and supporters of a policy for ensuring, in their collective perspective, democratic global communications. The documents listed in the UNESCO publication indexes pertained to several activities of the organization besides the work on an NWICO policy. Non-UN publications on the NWICO debates were also available, such as articles published in key academic and policy journals at this time.

Texts for the study have been selected from a bibliographical listing for sources pertaining to the NWICO published by the Dag Hammarskjöld Publications for the UNESCO. Documents pertaining specifically to communication and development in this listing were of particular interest for this study. Additionally, reports from the NAMEDIA publication and documents from the listings used for the MacBride Report were also included, along with selected academic journal articles published at the time on NWICO-development issues. To compile a list of pertinent documents for the study, key words used for identifying documents for the study included traditional media, community media and development, satellite communication, the more conventional mass media such as radio, communication, development, and social progress and social development. Approximately 20 official and 20 other texts, varying in topics and treatment were subjected to close reading. Illustrative texts from this pool have been analyzed to demonstrate the discourse at work.

There are two main tasks to this project. One is to understand the discursive strategies or mechanisms that went into the construction of certain notions of communication and development, and the other is to form an understanding of the place for development in the present form of globalization. I use tropological analysis to identify these discursive strategies. The detection of tropes is an interpretive project; as such, one needs to be conscious of re(constructing) a discourse. Zizek's (1993) consciousness of this reflexive quality to reading a discourse illustrates my point well:

> The aim of the discourse of the analyst always emerges for a brief moment: the aim of this discourse is precisely to produce the Master-signifier, that is to say, to render visible its "produced," artificial, contingent character. (p. 2)

That is to say the interpreter also participates in this production.

White (1978) called tropes the soul of discourse. If the logic of reality and the fantasy of fiction are seen as two ends of the continuum, White places tropes in the middle. That is, they are "prelogical" but not completely located within the realm of fiction. He explained this peculiar location within the context of historicizing society, where he identified the meeting point of historian and novelist, and where tropes and discourse can be identified.

Discourse works to limit the realm of meanings, understandings, and practice. White used terms such as *mark out, define, identify*, and *discern* to describe the labor that discourse must perform to retain the uppermost place in social consciousness. For the project on hand, this would translate to keeping a certain knowledge of communication and development uppermost in the collective consciousness of nations. To gain legitimacy, discourse must necessarily "[violate] the canon of logical consistency," and White termed this move as "more tropical than logical," thereby giving discourse an arbitrary character. Thus, some crucial ways in which the discourse of communication and development is established becomes accessible through tropes used to establish the discourse.

Tropological analysis has been used recently in colonial discourse involving extensive analysis of media texts. For example, Spurr (1993), in his work on the rhetoric of empire, developed a series of tropes to map the discourse of colonialism. Repetitions of certain themes formed patterns across time and texts and were articulated as tropes of that discourse. Similarly, Shohat and Stam (1994) employed tropes to study colonialism in Hollywood films and in the process, to demonstrate tropes as "an arena of contestation." For example, they demonstrated that race is "less a reality than a trope" and a category that is created through exaggeration and nuance (Shohat & Stam 1994, p. 137). In their analysis of Hollywood films, they demonstrated through tropes the play of what they called "schematic exaggerations," intended to consolidate the discourse of colonialism in film. Tropes are both "quasi-fictive" in nature and possess a kernel of the "real," and hence are especially suitable for demonstrating the operations of power to direct discourse.

CHAPTER SUMMARIES

In chapters 2 and 3, I demonstrate the emergence of the limited understandings or meanings that constituted the dominant discourse of communication and development. These limits to understanding are set by tropes, each seemingly contradictory pair in a chapter. I have identified four tropes—enumeration, macrospeak, surveillance, and invisibility. In chapter 2, I examine the pair—enumeration and macrospeak. By enumeration I mean an elaborate and detailed inventory available in the dominant discourse of development; diplomatic ritual, inventories of archives (e.g., previous agreements), communications media and infrastructure, and ways in which communication had to be organized and managed to achieve the state and status of development. Through this trope, communication and communications media are transformed into resources to be mobilized for development. The obverse of enumeration, "macrospeak," also contributes to the discourse to the same effect. I use the term *macrospeak* to connote a certain largeness and inclusiveness of address that coopted possible opposition. Hand in hand with this cooptation went a certain inclusiveness of whole regions in the rhetoric, regardless of significant differences among these regions.

I present the second contradictory pair of tropes in chapter 3—surveillance and invisibility. These tropes address Third World accountability for the use of communication in myriad ways to ultimately reach the stage of development. Three different strategies are discernible under the trope of surveillance. These strategies demonstrate the means by which power operates to secure the discourse of communication and development. First, especially apparent in the domain of satellite communications is the emphasis on rendering natural resources transparent (ostensibly) for governments to use in the process of development. Second, evaluations of existing and past projects in development communication were treated as lessons from which future projects (with modifications) had to be planned. A third manifestation of the trope of surveillance was apparent in another practice—Third World progress reports to the donor countries on communication projects, which would lend transparency and accountability to their efforts. Surveillance demanded Third World accountability for its communications activities in keeping with its position in the international arena—as a follower of the economically advanced western nations, if not as an identical copy, then in ways sufficiently recognizable to be termed developed.

In analyzing the trope of invisibility, in a sense the other of surveillance, I attempt to articulate the powers to which the Third World is accountable. The idea here is that the hidden overseers of development communication projects guide projects and decisions without explicitly stating their own role or presence. Additionally, economically powerful countries portray themselves and are sometimes treated by Third World countries as parent nations raising the Third World to full development (the metaphor of the parent–child relationship in the discourse of development can be found in the critical literature on development and colonialism). The parent nations thus possess the knowledge of how to deploy communications to fulfill certain political, economic, and socio-cultural criteria that they purportedly desire all nations to achieve. In this process, the presence of the power centers in the discourse of communication and development is mystified, as more of a shadowy entity working behind the scenes to effect Third World development through communication.

In chapter 4, I present two themes of negotiation and resistance, both of which emerge from the rhetoric on "another development" employed by many countries in the developing regions striving to maintain presence and a modicum of economic independence in the international arena, expressed mostly within the metaphors of development. In the theme of restructuration, drawn mainly from the nonaligned countries' remapping of world communications, I attempt to discern a shift of the center, to the periphery. For example, one such shift involved rethinking the use of satellite communications—from locating resources to mutually sharing among nonaligned states the experiences with problems and solutions associated with development efforts. In this proposed changed use for satellites, the interactions were confined to thinking of the nonaligned countries as a central region of collective concern, if not geographically, then extending Nandy's observation to the context of this study, as a community of the colonized in the master project for social change (Nandy, 1995).

Under the trope *retheorizing communication*, I read attempts to generate an understanding of communication that is markedly removed from the dominant discourse of communication in relation to development. Ideas such as communication as expression, rather than function, and communication as negotiation, rather than instrument, are two such examples of attempts to articulate another understanding of communications from and for the Third World. It lays the ground for the concluding chapter, where I bring together some recent attempts to go beyond the angst associated with membership in an asymmetrical world system, especially in the realm of communication and social change. I examine local media and establishment of identities, and communities' engagement with the local media to solve local problems that may or may not have originated at the global level.

In the fifth and sixth chapters, I address the shift from an international to a global era and the role of communication and development/postdevelopment in relation to this shift. Characterizing the global as a new phase of a recurring historical phenomenon (of which the international post-World War II would be one), I demonstrate the continuities and changes in the idea of communication and development in both eras that are currently the subject of much theorizing and analysis. Dismissals of the project of development in the global era notwithstanding, I demonstrate the continuity of the questions and the changes in the context which is now being termed postdevelopment. I examine the forms that development has taken with the use of new media technologies and discuss the implications of new conceptualizations of development in relation to power and democracy in the information age. In the process, I briefly address possibilities articulated by transnational development corporations and their use of the Internet, and new media technologies for social change and their implications for democratic communications through new forms of networking enabled by new media. These new ventures range far beyond the statist concept of development so central to the NWICO debates. In the sixth and final chapter, I attempt to place the role of civil society in the information age within the context of WSIS as a renewed quest for communicative democracy and development. Whether this move into globalization from below will make for a more equitable distribution in communicative power requires empirical answers and remains to be seen.

This project raises the question of whether indeed developing communication can bring about democratic communications in the international space. In a theoretical examination of the global public and political spaces, Venturelli posed this question from a policy perspective: whether communications technology, access, and delivery (read development) do indeed promote, ensure, and maintain a global public sphere (Venturelli, 1996). In the context of this project, we might modify the definition of communications technology to mean any form of communication mix or media in general that serve a development project and that reach the potential or intended publics. We might extend Venturelli's question to ask whether the NWICO was intended to establish democracy in the transnational public space or whether, through the complexities of the discourse, it served to reify the larger project of development, a signifier of inequality that denied the idea(l) of a global public sphere. We might further extend this question to speculate at least somewhat informatively about the discourse and outcomes of the WSIS as a next marker in the history of communicative democracy, development, and social change.

2

PROJECTS AND PROPOSALS FOR COMMUNICATION AND DEVELOPMENT

During the NWICO debates, communication projects, proposals, experiments, and interventions constituted the focus of many discussions that figured in symposia, were recorded as reports published by the United Nations and the UNESCO, and were analyzed in academic and other journals. Although these efforts were intended to demonstrate the viability of a new world communication order, what emerged simultaneously was an assessment of communications infrastructure and the need for further communications development in the Third World. In other words, the discourse of development paralleled the discourse of democracy in this instance. Third World development would correct imbalances in information flow and create the ideal environment for a democratic and free flow of communication with the establishment of sufficiently developed modern communication systems. Securing the dominant discourse on communication and development included taking stock of existing infrastructure and mandating what routes communication and development should take, how communication should and could contribute to development, and what was necessary to enable communication to set Third World societies on the course of development. Two major tropes emerge from a reading of projects, proposals, research reports, and reports of official meetings on communication and development—enumeration and what I have called here as macrospeak.

Enumeration involved listing, counting, and detailed descriptions of official procedures in meetings, projects, communication potential, and media applications (traditional, modern, and high technology) to foreground the need for communication's role in the larger enterprise of development. Enumeration also involved detailing the official, dominant, viewpoints and past development project experiences. Conversely, macrospeak refers to all-inclusive, generalized discussions of communication for development often involving normative statements. Repeated assertions of very general and normative statements resulted in big pictures that often alluded to ideal modern societies, missing a level of specificity that might have contained the potential to address problems like the serious lack of opportunity structures, internal conflicts with varied histories, and multiple public opinions that might have helped germinate alternate viewpoints and solutions. Both tropes constituted part of the dominant discourse of communication and development.

Third World governments and their representatives at international development fora found themselves in the precarious position of having to constantly negotiate a tension between maintaining good relations with donor constituencies in the advanced economies for development aid, and their need for changing the nature of the existing communication order to a more balanced one suited to their respective needs and circumstances. Development, then, needed the benevolence of the economically powerful countries to recognize the importance of a balanced flow that would help with development rather than continue to support the free-flow philosophy. Many in the advanced economies objected to what they perceived a balanced flow of information would entail—a state-controlled flow that, in their view, would impinge heavily on freedom of expression (Kang, 1988). Many documents read for this study, particularly official texts, demonstrate this tension where Third World nations participated in the dominant discourse of communication and development, yet also maintained contestatory positions that emerged within the legible metaphors of the discourse.

Texts read for the two tropes in this chapter are located at intersections—of the international (e.g., studies commissioned by the UNESCO) and the national (representation of traditional media cultures), and of the desire to preserve tradition and the desire to achieve the state of modernity (which included improving health and nutrition, education, gender equity, human rights, and economic self-sufficiency). Often these intersections characterize the developing nations' desires and aspirations at a national level. On the one hand, there is deep knowledge of and appreciation for traditional, non-Western media arts that occupy dual spaces as high culture at the level of the state and popular culture at the level of the people, particularly in nonurban areas. On the other hand, in the sharing of this knowledge for a particular purpose, the entry points for utilizing these media for development were explicitly identified from theoretical perspectives specific to media studies; utilizing these entry points for development projects suggested a move away from communication as popular expressive culture to an instrumental rationality that sought "maximally effective ways of organizing or educating people" (Chouliaraki & Fairclough, 1999, p. 12), predicated on more limiting meanings of development. These intersections explain the nature of some of the documents under study here—one of cooptation into the dominant discourse on development struggling with a more indigenous voice. At the helm was the UNESCO, itself in an ambivalent position. The UNESCO's empathy toward developing regions, under the directorship of Amatar M'Bow, and its agenda for working with them helped materialize many of these studies. But the internal pressures exerted by advanced nations within compelled the organization to produce a softened rhetoric that in the end supported technology, economics, and Western definitions of development and democracy. Expectedly, some of the texts read for tropes of domination displayed certain tensions, notable among them those that were commissioned by the UNESCO. In occupying this dual position, these texts contribute to both the dominant and resistant/negotiatory tropes. I have created (an albeit artificial) separation of the tropes and have discussed some texts under both sets of tropes.

The tropes in this chapter—enumeration and macrospeak—allow us to see how the discourse of communication and development was constructed within certain bounds and ways in which they "[legitimated] an arbitrary boundary" (Bourdieu 1991, p. 118) between the developed and the developing countries, emphasizing the lacks of one in a constant play against the other. Official domains such as the UNO and the UNESCO operate on both constructed and received traditions of formal discourse. They constitute a site where the "authority underlying the performative efficacy of discourse" (Bourdieu, 1991, p. 106) is established. The "rites of [diplomatic] institution" work to establish an "authorized way of seeing the social world."

Communication projects and proposals for modernization and development constituted a major topic of official discourse. In the context of the NWICO debates, the idea was to accelerate development of Third World countries to enable them to generate sufficient quantities of information and communication on their own, which would ostensibly correct the existing imbalances in international information flow. Therefore, the attention given to communication for development in this period is particularly noteworthy. Official documents represent what sociologists Burton and Carlen (1979) described as "a system of intellectual collusion whereby selected... intelligentsia transmit forms of knowledge into political practices" (p. 15). One discursive strategy employed in official documents during the NWICO debates was that of containing communication within certain functional roles. Out of the three basic roles typically identified for the media, the tendency was to emphasize the education function, rather than the information and entertainment functions. Information and entertainment have been integral to traditional media in many Third World cultures as a parallel to the modern media in the developed world. In many cases, entertainment media forms familiar to nonurban and peripheral urban audiences were identified as vehicles for development messages.

Finally, Fair, (1996), in her work on media coverage of the famine in the eastern horn of Africa, described news production/knowledge production in an androcentric Western media environment that is particularly damaging to the women in East Africa. Her description of the process is helpful for understanding the construction of the discourse of communication and development as well. The following quote demonstrates the normalization of the discourse that helped produce formalized types of, to use Bourdieu's term, *habitus* such as media institutions:

> The ideological forces of relations of ruling is that the social organization of knowledge serves to subdue and dislocate the perspectives of lived experiences, replacing them with categories, facts, rules, and policies set by the organization producing the texts. For this ideological process to have success organizations must be able to construct and structure reality in ways that make relations of ruling operating within the text invisible (Fair, 1996, p.6).

Fair demonstrated these dislocations empirically by bringing to the center a population at the margin. By critically examining representations of women in the horn of Africa in U.S. TV news during the Ethiopian and Somalian famines, she described the ways in which these women were reduced to little more than powerless wasted "bodies," an unfortunate representational outcome of androcentric Western media frames, and one that, in other regions and with other disasters, reproduced the dominant representations of developing regions. The history of representations of marginalized populations in the coverage of international news also contributed to the emergence of the dominant discourse. Such representations emerge from certain definitions of news that I examine in the trope of enumeration.

In the UNESCO documents in particular, the establishment of "relations of ruling" (Fair, 1996) occurs at two levels. At one level, government representatives participating in the arena of diplomacy are organizers of social knowledge for their respective peoples. At another level, within this diplomatic site, the more powerful nations organize knowledge for all nations. As Alvares (1992) reminded us, power decides what is knowledge. Discussions at diplomatic/official levels on treaties for communication and development, projects, agreements, codes, and policies served to emphasize ways to see how communication could be harnessed as a tool for development, or for educating underdeveloped communities, emphasizing education over information and entertainment.[1] As we will see in the tropological analysis, these discussions circumscribed the domain in which communication and development could be interpreted. Thus, communication was treated more as an

---

[1]The concepts of "enter-educate" and "edutainment" were introduced more recently in approaches to development with the media.

Projects and Proposals 25

instrument for attaining development of a certain kind, derived from a certain history, and less as a source of cultural expression for multiple reasons and occasions that could exist outside this body of meanings. Consequently, the dominant meanings of communication and development attempt to contain broader interpretations of the concept. I demonstrate ways in which the tropes of enumeration and macrospeak worked to define the range of meanings of communication and development.

ENUMERATION: TAKING STOCK AND SPELLING OUT STRATEGIES

Many examples of the trope of enumeration as a mode of establishing the discourse of communication and development through details of plans, procedures, and strategies are apparent in certain texts. Three modes of enumeration in particular are worth noting (see Table 2.1). *First*, in official publications of reports of meetings, the protocol and procedures of each meeting were recalled in detail. Also, frequent references were made to past documents, reports, and agreements on communications. This material was drawn from archives for presentation at meetings, conferences, and symposia on information and communication. *Second*, existing and potential communications and information infrastructure, both traditional and modern, were inventoried in some detail in reports, studies, and journals. Stock taking served to set the stage for discussing how communication would help

TABLE 2.1. ENUMERATION

Methods of Enumeration	Themes
Official rituals and archives	• References to and restatement of past meetings, documents, and agreements.
	• Institutional protocol.
Delineating infrastructures and institutions	• Technology and its capabilities.
	• Communications media as development resources.
	• Development potential in traditional media.
Organizing and managing communication for development	• Management of communication technology.
	• Magnitude and urgency of the need for communications technology development.
	• Journalistic practices.

achieve development, and what was needed to make developing regions satisfactory producers of information for the world information flow rather than remain primarily as consumers. In the process, cultural practices were converted into resources to be used for economic development.[2] The assumption underlying this conversion was that governments should think fundamentally about educating people to "become developed," and that for this purpose all communicative resources at hand and beyond (considerably beyond—hence development aid) had to be pressed into service. Third, many texts dealt with the organization and management of communication for development. Elaborate plans, analysis of past experiences, and experiments on the use of satellite technology or traditional media for the purposes of development reemphasized certain ways of thinking and speaking about a specified set of roles for communication to achieve development.

Rituals and Archives

Official reports elaborate the protocol employed in official meetings and, in the process, this documentation serves to reproduce the official discourse. As forms of accountability for meetings and documenting related procedures, official reports require a special rhetoric to elaborate protocol that results in (re)producing that discourse. Recalling past agreements among the producers of this discourse is one such practice that enables its further consolidation. Phrases such as "agreeing with," "recalling that," "honoring," and "noting," followed by titles or numbers of official documents and reports perform this protocol. The 107th Plenary Session of the UN General Assembly is an example of this genre of official discourse.[3] A section of the report entitled "Co-operation and Assistance in the Application and Improvement of National Information and Mass Communication Systems for Social Progress and Development" begins with, "The General Assembly,

---

[2]Most of the critical literature on development points to economic development as the emphasis in development projects. Economic development sets the parameters within which all other types of development—social, cultural, political—have to be interpreted.

[3]See, for example, the Report on the United Nations Economic Commission for Africa Regional Seminar on Remote Sensing Applications and Satellite Communications for Education and Development (United Nations, 1981). See also the Report on the United Nations Regional Seminar on Remote Sensing Applications and Satellite Communications for Education and Development (United Nations, 1981). Similar examples can also be found in a subsection of the plenary meeting report: "Concerns about cooperation and assistance in the application of and improvement of national information and mass communication systems for social progress and development."

Recalling its resolutions 1778 (XVII) of 7 December 1962, ... of 1973, ... of 1976, and ... of 1978, etc." (United Nations, 1979—1980, A/34/46, p. 82). Similarly, in the next section of this document entitled "Questions Relating to Information," several articles, declarations, and documents were invoked. The report enumerated diplomatic protocol in the organization of the minutes of the meeting, in its several references to previous meetings on the issue, and to other related reports and agreements. References to earlier agreements and documents served to maintain the continuity of the rhetoric in which the discourse of communication and development, to a large extent, was embedded. The annex to the 107th Plenary Session of the UN General Assembly on communication, social progress, and development elaborated procedures for institutionalizing communications and related activities for development. A proposal was called for the "institutional arrangements to systematize collaborative consultation on communication development activities, [and] needs and plans" (United Nations 1979, A/34/148, p. 1). The institutional role and the act of systematizing at this level contributed to defining dominant designations of the role of communication for development.

Anthropologist Bloch argued that such official rituals have a defined effect on the discourse they produce. Equal emphasis is laid on "what *can* be said, [as well as] the *manner* in which it can be said" (Bloch, 1975, p. 5, italics original).[4] The modes of expression can restrict the ways of thinking about the discourse. As Bloch (1975) pointed out, "this type of restriction is...more powerful than a direct attack on content, since it goes right through the whole range of possible responses" (p. 5). Together, rituals and archiving practices followed by these institutions served to establish certain ways of thinking about communication and development.

Infrastructures and Institutions

Detailing existing information and communications media and infrastructure, and logically extending the descriptions to what would further be needed to reach the state of development, was another discursive pattern apparent in this trope. In a UNESCO document entitled "The Philosophy of Development Communications: The View From India," the author recommended propelling developing societies "along the path of modernization." Communication acquired its meaning from its ties to (modern) tech-

---

[4]Anthropologist Maurice Bloch studied oratory in a ritual/formal setting among the Merina of Madagascar. My analysis draws on his astute observation of formal discourse, where certain modes of address and references characteristic of traditional oratory societies carry to formal and ritualistic discourse apparent in documents recorded in modern formal and ritualistic settings of suprastate organizations.

nology; "Since communications is both an instrument and a product of development and technological advancement, it is no surprise that developing societies are disadvantaged in this regard" (Verghese, 1978, p. 1). Within this context, this statement locates communications in the domain of technological development. The document also details existing and potentially needed policy, management, and infrastructure in developing countries. For example, post and telegraph services, a telephone system, satellite communications, and electronic data processing would be harnessed for development planning. Following a documentation of rural broadcasting in India, an extension setup for information transfer in the agricultural sector, the document concluded that India would have to continue to rely on a combination of traditional and modern media for development purposes for some time to come (Verghese, 1978). This prescription invites another reading—that new technology and mass communication would eventually become predominant modes of communication, and the continued co-existence of alternative communications could diminish. The specific spatial arrangements advised in this document resonate at times with Lerner's ideal modern society—in the envisioning of more modernized cities and towns. In general, the document detailed communication achievements in India; specifically, it highlighted the shortcomings. As part of the solutions to overcome these inadequacies, this document envisaged a mix of traditional and modern media at first that could later lead to a fully modernized media system.

In another UNESCO report (Gazin, 1978), an argument was advanced for satellite communications and broadcasting as the most effective medium for development. This document serves as an example of the trope of enumeration in the context of discussions about satellite communication technology. Elaborating on the differences in communications capacities and capabilities between developed and developing countries, such as speed of communication, message-carrying capacity, immediacy, cost, and resources for maintaining technologically advanced communications systems, this report participated in the dominant discourse in two ways. It delineated differences among groups of nations based on communication technologies, and omitted to address the history behind the birth of the technology and the general ideology behind the power invested in such technologies. The report highlighted the economic side to technology, constructing differences in development between sets of nations thereby promoting the need for developing communication technologies. Further, the report predicted that "it can here and now be stated with certainty that the 1980s will experience the most explosive growth rate . . . of orbital means of communication, transmission, [and] exchange," recognizing technologically advanced media as the signifiers of growth, development, and innovation (Gazin, 1978, p. 13). In the process, the report did not specify that these media may indicate one of several types of growth for certain types of societies.

With regard to satellite technology and orbital space allocation, the discussion in this document included transmission/broadcasting beyond national borders, assurance of high-quality reception, and cost of receivers. The document described the power and potential of technology, but did not raise the issue of power differentials inherent in the exploitation of technology and the implications for the crossing of a significant territorial/political marker—national boundaries. Together, these two approaches ran the risk of keeping the orientation of the report to conventional definitions of development.

Finally, the document provided elaborate solutions for the problem of overcrowding in satellite stationing space, such as the "platform approach," where several satellites would be stationed at approximately the same orbital space to enable broadcasting to a larger number of countries. These solutions to satellite overcrowding in outer space raise several other questions around the ownership of satellites, setting the price of information, the clientele for the information, and where the profits would ultimately accrue. A general idea of cost feasibility for the supplier was discussed in the report. Here, promoting a strategy involving large-scale technologizing of communication in developing regions subordinates ideological implications. This is apparent in statements such as "scientific and technological development in the fields of communication and information already offers, and will continue to offer in the era which we are about to enter, an exceptional opportunity . . . successfully to fill the gap between developed and developing countries" (Gazin, 1978, p. 18).

In an article on planning communications for less developed countries, Cleevely and Walsham (1980) argued that telephones were an indispensable medium for rural development. They held that telephonic communication would contribute substantially toward alleviating poverty in the less developed countries. Drawing links between telecommunications development and economic development, this article illustrated the importance of favorable and enhanced investment and pricing policies for telecommunications in the overall picture of public investment policy. To understand how telecommunications investment and pricing would have to be treated in overall development policy, elaborate models were presented to build the idea that high infrastructure signified high economic development, which would ultimately fetch a high telephone revenue per line. Also, the article emphasized the importance of assigning a unit price value to a telephone call to render it manageable in economic, and hence public policy investment terms. The marginal social product of an extra unit of resource, or the social value of a telephone call, needed to be priced especially because it is a "non-traded" good. In other words, human contact over the telephone (around which idea, ironically, U.S. telephone companies build their USP to market their services—e.g., the telecommunication giant AT&T's Reach Out and Touch Someone campaign in the late 1990s) is transformed into a non-traded good to be factored in financial calculations as "shadow pricing," and thereby to feature in investment and pricing policy for development.

This against-the-grain reading of a text on telephone technology rests on certain assumptions and points to limits of the dominant development discourse: (a) that all societies eventually need high levels of telephone use and infrastructure; (b) that diverse structures for telecommunications other than the economic are perhaps secondary in importance; and (c) that development of rural areas depends on the availability and use of telecommunications facilities. This is not to deny the manifold conveniences that telephone communication offers, but the underlying ideology is familiar; the text illustrates another communications domain where detailing "environment," pricing, investment, and policy in relation to telecommunications—all in the economistic realm—emerge from the dominant discourse.

Like new technologies and modern media, traditional media and their uses for development were also enumerated in detail. Traditional media are critical for reaching rural publics for important messages, yet their didactic use for development could color cultural practices and their own rhythms of change, transforming cultural expressiveness into a vehicle for development messages. Traditional media were absorbed in the broad discourse of development. Their independently constitutive cultural role as "expressive traditions" became (at least temporarily) secondary, and the "tradition" component in the practice of these types of communication was alternately recognized as safeguarding (pure) cultural tradition or integrated into the development discourse.[5] Both history and studies of popular culture have shown that folk media, besides constituting expressive cultural practices, have long served as vehicles for information, political persuasion, and social commentary (this is evident in works with diverse theoretical and philosophical orientations. For example, Malik (1981), and Spivak (1996), in two very different instances, empirically demonstrated these roles for traditional media. Malik's work is discussed later. Spivak explicated the role of street theater in Bengal for spreading the ideology of nationalism). For example, two documents on India, referred in the following discussion as Documents 1 and 2, and another report on Egypt, detail the structural openings available within traditional media for the insertion of development messages designed to reach rural publics. These documents contain excellent and concise descriptions of a wide variety of traditional and folk media and theorize traditional media from non-western perspectives. Beyond description and theorizing, the documents identify opportunities for or the ways in which the development idea can be inserted in the message. An Indian folk media specialist is quoted in Document 1:

---

[5] I borrow the term expressive cultural traditions from Appadurai, Korom, and Mills (1991).

> ... the rural drama encompasses almost the entire inner personality of the villager. It seeks to meet all his intellectual, emotional and aesthetic needs. Unlike urban and modern drama, it freely uses songs, dances and instrumental music, besides dialogues. This multiple approach results in a form that is self-contained and complete entertainment for the audience to whom it is directed. It is more than an entertainment; it is a complete emotional experience and aims at creating an environment of receptivity in which communication of ideas is an effortless process. (Ranganath, 1979, p. 2).

This document refers to a division of the state in India—the Song and Drama Division—established primarily to support folk arts and traditional media forms, and describes the incorporation of traditional media for development purposes. The Song and Drama Division occupies a paradoxical position in the discourse of communication and development in India. It is charged with safeguarding traditional cultures and works to preserve traditional media in their more pristine forms. Yet, by extracting the uniqueness and characteristics of these media to manage them for development purposes, the state enables the use of traditional communication as a tool for development. To use the analogy in the document, it would be as effective as "us[ing] a nail to drive out a nail" to drive out tradition (in this context, underdevelopment) with traditional media. Undeniably, important messages in certain domains like health (with origins in both Western and non-Western medicine) are conveyed effectively to rural populations, and folk media indeed do represent an important vehicle in this context. Social messages designed to make the public conscious of gender inequality find a place in traditional media communication. However, the didactic tone of modernization in this context competes with the cultural experience derived from folk media performances. The analysis concludes—traditional media have "revealed some secrets of getting the best out of them with the least hurt." This extraction of the best of the traditional media to serve the development project is also demonstrated in Document 2 discussed next.

In an analysis of traditional media in India, Document 2 observed that art and education were integrated historically in India because of the recognition of the social role of art (Malik, 1983). An indigenous communication theory, the *rasa Siddhanta*, which explained the complete absorption of audience members in folk theater to the point where they would become highly receptive to messages the performers conveyed, is effectively developed in this document. In effect, Natyashastra, India's "earliest manual of dramaturgy ... teaches duty ..., promotes self-restraint, ... gives courage ... energy, ... [and] wisdom" (Malik, 1983, p. 29). Perhaps one of the earliest examples of the entertainment/education approach in development communication today, the cultural form of folk theater presumably served the needs integral to community living more than transmitting values and attitudes suitable to the ideology of development. Document 2 on traditional media in India suggests that this characteristic of the traditional media, combined with the theoretical opening in folk theater, can be appropriated for positioning development messages.

One other document on traditional media in Egypt (1980) with similar descriptions bears a close reading. This document described in detail traditional forms of communication in Egypt using the vocabulary of modern communications media. It first identified some issues that traditional media share with more modern mass media such as print or broadcast, and with related practices such as advertising but with a different emphasis. The emphasis was on source credibility (high regard for religious leaders who preside over many religion-based folk and rural media), and on opportunities for inserting subliminal messages in the folk form like the *Djar*, a form with the power to induce mind-altering states through ritual group chanting.[6] Suggested locations for use include government service arenas of folk and other venues such as the cultural courts. For the effective use of traditional Egyptian media for development, systematic planning and other strategies derived from modern media activities were recommended, in scale (nationwide) and locus of control (the national and state development organs). Activities involved identifying strategic target audiences, situational assessment of existing knowledge and attitudes, and a system for "continuous measuring of the results of different communication systems and for increasing their effects" (Hussein, 1980, p. 7).

These documents are valuable for the specialized knowledge they contain about traditional media and the sensitivity with which these media have been treated as communicative and expressive traditions. The broad notion of development and development messages contained in these texts places them in an ambivalent position in relation to the ideology of development because of the nature of the development discourse. Incorporating traditional media into the developmental fold helps transform them from cultural forms constituting part of a social fabric into resources and vehicles for a certain type of social change, with defined functions and uses toward this end. This form of development is based on a vision of a changed society that is universalized to the extent of adapting all possible cultural resources to reach that ideal destination. There is also a danger of freezing cultural pasts, when in fact we can reasonably assume (and as document 1 has noted) that as cultures change so can (and do) the traditional media.

---

[6]Although advertising research has shown that subliminal messages have proved to have little or no effect, in the context of this document, it is not the message technique, but the intention to use this technique for development purposes that transforms this idea into a development strategy.

Organization and Management of Communication for Development

The stock-taking discussed in the previous section served to compile a knowledge base of available communications means and sources for development projects. Organization and management of this knowledge base for effective use of communications for development also constitutes the trope of enumeration. Here the tasks, institutions, modes of executing tasks, materializing plans, and detailing operations were directed toward using communications resources to close the gap between the information-rich and information-poor countries. One domain in which the creation, emphasis, and consolidation of this idea of the gap, and its role as an index of development/underdevelopment, were apparent was in satellite communications and related technology. The organization and management of communications for development served to first emphasize the absolute and unquestionable need for satellite technology as a vehicle for development. The scope of such management and direction was then established in official reports such as the ones on the seminars on remote sensing and brought to attention in critical project evaluations such as the report on the SACI/EXERN project in Brazil carried out in the decade between the mid-1960s and the mid-1970s. The satellite seminar thus constructed a serious need for aspirations to reach maximum communication capacities through satellite technologies. The centering power of this text emerges from this reading—once the advocates of satellite communications emphasized the need for satellite technology worldwide, warned of the dire consequences for the laggard nations, and elaborated a series of management solutions to install (Africa seminar) or improve systems (Brazil, India), inevitably, the power would lie with select bodies of decision makers to direct communications for development.

Between 1967-1974, the SACI/EXERN project was implemented in rural northeastern Brazil for the use of satellite technology and broadcast media in education. In a critical evaluation of communications planning and management of this project, McAnany and Oliveira (1980) observed that Brazilian central planning in the SACI project suppressed initiative of the individual states. The rationale for executing such a large project was its cost-effectiveness, which could not be achieved through either traditional schooling or individual state initiatives. In this thinking, elaborate planning acquired considerable importance:

> A tight coupling between the managerial and the technical approaches was . . . obtained [through the systems approach]. Behavioural objectives, deadlines, flow charts, formative evaluation, cost effectiveness, rational plans—these are some examples of the language of planning at INPE headquarters. (McAnany & Oliveira, 1980, p. 19)

The authors also pointed out that the intense focus on planning and the systems approach to managing satellite communications for education limited the planners' resources and resourcefulness for dealing with unplanned outcomes. The "rational model of management and education" provided the engine for the planning. Some of the problems in this project were attributed to factors often found within the discourse of communication and development—the "natural difficulties arising from the backward situation of teachers, teaching and students in the state" was one such problem—teachers and teaching were not sufficiently technologically and nationally disposed to accept a system such as the one set by SACI (McAnany & Oliveira, 1980, p. 35). Another was the "lack of adequate expertise with the techniques embodied in the [systems] concept," again a problem of inadequate training of personnel in a particular (technological) domain (McAnany & Oliveira, 1980, p. 35). The project's *raison d'être*, its cost-effectiveness, could not be demonstrated adequately because of lack of support from education or communication establishments.

Although the experiment was evaluated as a novel attempt in education with potential for the future, questions relating to content and pedagogy, and the central authority's (the INPE's) management style were thought to limit the success of the project. Concluding their critical evaluation of the SACI EXERN project, McAnany and Oliveira (1980) posed a perceptive question about a technological agenda—whether acquisition of a domestic satellite for Brazil in the foreseeable future might actually hamper the success of the project as originally envisaged by its planners. This question implicitly points to the centering of technology characteristic of the dominant development discourse at the time.

In a Regional Seminar on Remote Sensing Applications and Satellite Communications for Education and Development in Addis Ababa, Ethiopia, considerable attention was given to organizing and managing satellite communications for development. The participants in the seminar on satellite communications in Africa advocated the help of the United Nations in disseminating information to assist countries in formulating communication needs. In this sense, management was not confined to communications alone, but also to directing what the communications needs of a society should be. Technologically advanced communications then become a natural activity in which all societies were expected to participate. Elaborate instructions on the use of surveys along with an "inventory of resources" were illustrated with flow charts indicating the sequence of activities (United Nations, Addis Ababa, 1981, A/AC.105/290). Seminar participants advised African countries to establish a National Remote Sensing Center for specific development-related activities for optimum use of satellite communications. The centers were advised to "initiate . . . implement . . . monitor . . . set technical standards . . . recommend procedures . . . prepare manuals, design courses . . .

[etc.]" (United Nations, Addis Ababa, 1981, p. 12), all of which would serve to harness communications from a central institution in each country and employ it as needed for education and development purposes. Development was thus imagined in centrally controlled terms. The rhetoric of control and management served to set the stage for some (indefinite?) point in the future when African countries would use satellite communications as a matter of course (that is, reach a true state of development). Specifying and enforcing the use of certain communication technologies in certain ways thus circumscribed the discourse of communication and development.

A similar conference on Remote Sensing Applications and Satellite Communications was held in Buenos Aires earlier that year (1981). Many issues raised in the Addis Ababa conference appeared in this conference also, with the potential for future use being the focus of the Buenos Aires seminar. "Efficient planning of resources" required "accurate, reliable, continuous and timely information" for development (United Nations, 1981, Buenos Aires, A/AC.105/290, p. 4). Information became a system "basically comprising data collection, information extraction and information feedback." Thus, the language of satellite technology defined communication as a linear progression of information use and feedback; control over this linear system would enable state powers to act out, in part, a certain vision of development through the vehicle of satellite communications. Similar to the previous conference, in keeping with the language of development, here too primacy was given to the technology, and experts suggested shaping human resources to fit technology through specific kinds of training, even for communications needs identification because the technology was "created ahead of the user's capability to specify his requirements" (United Nations, 1981, Buenos Aires, A/AC.105/290, p. 5). Thus a communications medium born out of another ideology (that fueled the information revolution of the developed world), needs (Western military), imagination, and history (European) came to appear as the natural course of a world development history. In the process, space was transformed into a scarce communications resource, and ways to manage the space efficiently for stationing satellites also required elaboration.

The organization and management of communication for development spanned a wide range of media and media activities in a report on a conference on Consequences for Development Policy held in Bonn in 1978.[7] Participants included journalists from Third World countries, media policymakers, media experts, and members of international and regional organizations. Print media, planning, basic and advanced journalistic training, and access to the developed countries' media markets (through exchange programs, "fact-finding travel," etc.) were some areas for which recommenda-

---

[7]See Bielenstein (1978). I provide a close reading of the last article of this conference report. For similar arguments, but in a different context, see Nawaz (1983).

tions were made. A central issue in the NWICO debates—imbalances in news exchange between the North and the South—was addressed in some detail. The working group acknowledged that including alternative news about the Third World "[would] depend on the willingness and ability of the industrial countries to take up such alternative offerings." One of the questions raised included editors' willingness in the industrialized countries to accommodate development news adequately, given that the alternate nature of such news marginalizes it from the industrialized countries' mainstream media interests. To this end, the group recommended training First World journalists and gatekeepers such as editors to recognize the importance of disseminating alternate or development news. But the "history of developing expectations from the media," as Ogan and Fair(1984), citing Abel, have pointed out, had prompted the public to subscribe to the media industries' definitions of hard and soft news, where hard news has a history of selling, and alternate news is seldom hard news.

Therefore, it was recommended that Third World media professionals study the international information market—"It was conceded that in the North there is an 'information market' which the Third World countries must try to enter with market-conforming means" (Bielenstein, 1978, p. 110). As the recommendation for the journalists of developing countries reads, if the market continues to dominate, then it is hard to imagine editorial departures from routine practices, where advertising and audience demands are important factors for deciding content. Hence, this recommendation did not represent a radical departure from the status quo.

Additionally, it was also suggested that journalists from developing countries be assigned for specific periods to the newsrooms of media in industrial countries. These measures—whether the reeducation of the First World journalist or the visiting programs for the Third World journalist—accepted the existing definitions of news, which, through journalistic conventions and practice, have the power to "arbitrate knowledge" and define the boundaries for public discourse (Tuchman, 1978, p. 14). The implications of such definitions for public opinion building have formed the subject of important works in the sociology of news. Such practices (such as single reports, short journalistic observations, spot news, breaking news, and, in general, inadequate coverage) that contribute to setting parameters for perceptions and meanings become detrimental to reporting a process like development as a complex event. Tunstall (1977) observed that "the media occupations are pervaded by the art of the anecdote; journalists are professional anecdotalists" (p. 37). The U.S. media baron Hearst's achievement as "tak[ing] the American western tradition of the tall tale and dress[ing] it in clothes of fact" (Tunstall, 1977) serves to emphasize the anecdotal nature of reporting and the narrative qualities of news.

Projects and Proposals 37

To change the treatment of development news in the newsroom would mean altering the definitions of news. Even if news is "event-oriented, timely . . . a change in something [that] can be dated and specified" (Ogan & Fair 1984, p.174), redefining *event* to include a more complex and longer duration, recognizing slower and less easily visible processes like development as change, and expanding the perception of timeliness to include an adequately responsive documentation of social change in developing regions would be needed to contest the discourse in the domain of development journalism (Schramm & Atwood; cited in Ogan & Fair, 1984, p. 174).

"MACROSPEAK": THE BIG PICTURE

I have used the term *macrospeak* to define and describe a second discursive strategy that involves a broad, inclusive, and normativizing treatment of communication for development. Macrospeak is differentiated from the earlier trope in that enumeration involved articulating detailed plans to help materialize the development role mandated for communications media. Macrospeak is characterized by elisions and omissions, where detailing events, procedures, and populations that could hold the potential to stimulate oppositional discourses was marginalized or coopted into the dominant discourse of communication and development.

Several expectations were tied to funding and aiding communications projects for development. The expectations ranged from expression of global statements of growth and development to inclusion of numerous communicative practices in the service of development. A normative tone is apparent in this trope in that the actors at the state and international levels would address how governments and institutions should function and cooperate in aided/funded development efforts, how a modernized society can be best imagined through the development of communication technologies, and how these technologies were necessary for development in other areas for broader social advancement. Referring back to Foucault's (1978) notion of discourse, the modes of utterance and enunciation in this trope favored a big picture, which directed attention away from substantive issues for social change in the Third World to a development discourse of technology and economics (Sachs, 1992).

I discuss this trope in two sections (see Table 2.2). In the first, I examine general statements about development. Addressing regions and continents as single entities, thereby discounting the intraregional differences, is also indicative of this generality. Especially in official documents, this generality is a result of the formalization of discourse through rules and protocol, themselves structuring principles of the discourse. Formalization directly

TABLE 2.2. MACROSPEAK

Methods	Themes
Generalized rhetoric and inclusive address	Absence of substantive discussions of terms like social change, social progress, and social development.
	Downplaying intraregional differences; treating the Third World as single entity for communications solutions to development problems.
	Normative tone for developing regions' communication activities.
Coopting participatory and expressive cultural practices	Absorption of Freirian conscientization by the state.
	Absorption of traditional media into the developmental fold.
	Absorption of grassroots practices such as participation into development projects initiated by state and suprastate agencies.

affects content. The content becomes predictable, is generalized, and in danger of losing "particularity and specificity," and therefore a historicity (Bloch, 1975, p. 13) that can provide a context from which critical questions about communication and development can emerge. The discourse of communication and development then renders all other types of society and lifestyles as alternative at best. This discursive strategy of generalizing tends to overlook the history of a particular kind of development of the First World, and relies on this development as the basis for prescriptions for Third World communications. In addition to generalized statements, normative and ideal goals as to how Third World nations should reach a certain stage of social existence and how the communications media should be used to this effect are also apparent in this trope. A didactic tone works in concert with the strategies of generalization and cooptation to hold Third World nations accountable for their use of communications for development.

Second, macrospeak is also manifested in the treatment of the "indigenous" — traditional media and grassroots communications activities rooted more in a traditional "sociality," such as folk and oral media and participatory communication. These media were absorbed into the dominant discourse.

Generalized Rhetoric and Inclusive Address

One manifestation of the trope of *macrospeak*, particularly in official documents, was a repeated reference to communication and development in broad and general terms. *Social change, social progress, social development*, and *cultural development* were used in the conclusion to official seminar reports and speeches where, in many instances, the main body of the report/speech constituted an elaboration of diplomatic protocol. Minimal substantive debates around these terms make it difficult to gauge the decisions taken about these issues. For example, the 107th Plenary Meeting of the UN General Assembly was called to discuss the "Application and/or improvement of national information and mass communication systems for social progress and development." The report indicates that the meeting was held primarily to confirm consensus on the use of communications media for development, although the title might suggest a more substantive discussion on how the media were being used and where their future use might lead. This macrolevel reference to communications and development is partly a function of its appearance in an official report of a meeting. In a discursive locale such as documented formal institutional practice, a specific vocabulary associated with ritual and formalized discourse, (Bloch, 1975) thus becomes apparent. Similar examples include a report on the Cooperation and Assistance in the Application of National Information Systems that advised "provision to developing countries of technological and other means for the purpose of promoting a free flow and wider and better balanced exchange of information of all kinds" (United Nations, 1979, 80, A/34/148, p. 1).

The larger threat of a cut in financial, technological, and professional assistance from certain First World member nations of the UNESCO led Third World nations to dilute the initial, explicit demand for a balanced flow of global information and communication and obliged them to include the dominant interpretation of a free flow of communication, and to list it in documents and reports first, followed by "balanced exchange" ("free and balanced"). Official discourse thus becomes a signifying practice—the larger idea of communication for development is central to the rhetoric, and official discourse becomes a "technology of ideological closure" (Burton & Carlen, 1979, p. 13). The report called for "us[ing] [communication] appropriately to enhance further economic and social progress of developing countries" (United Nations, 1979–1980, A/34/46, p. 83). As development theorists and activists have shown, enhancing economic and social progress within the definitions of the dominant development discourse and using technology for economic development acquired considerable conviction over time.

This discursive strategy appears in places in the MacBride Report. The Report was the result of the research efforts of the International

Commission for the Study of Communication Problems. The 11-member Commission headed by the Irish ambassador to the UNESCO, Sean MacBride, involved an international group of communications researchers and media practitioners in its research for the Report. The Commission compiled an exhaustive discussion of communication particularly in developing regions, the origins and histories of communication, communications technology and economics, and, finally, a policy proposal in the volume entitled *Many Voices, One World*. After a discursive battle of nearly a decade in the UNESCO, the NWICO proposal at the end of this report reflected an uneasy compromise with developed countries' demands for a free flow of information and communication. Generalized rhetoric was a strategy that signified cooptation of the NWICO demands by the dominant discourse of development, which existed in tension with the need to persist with the demand for a NWICO. The report recommended that the following considerations be included:

> The formulation of communication policies should: (a) serve to marshal national resources; (b) strengthen the coordination of existing or planned infrastructures; (c) facilitate rational choices with regard to means; (d) help to satisfy the needs of the most disadvantaged and to eliminate the most flagrant imbalances; [3] (sic) emphasize universal and continuing education; (f) help in strengthening cultural identity and national independence; (g) enable all countries and all cultures to play a more prominent role on the international scene. (MacBride Report, 1980, p. 207)

This quotation illustrates the more general nature of statements in the special report that rendered the demands for a NWICO as nonconfrontational as possible in a UNESCO climate that was rapidly turning hostile due to objections to a new information order from Britain, and from the United States under the Reagan presidency and its strategy of confrontation, a change from the strategy of accommodation adopted earlier during the Carter presidency (Kang, 1988).

Finally, one last instance of generalized rhetoric characteristic of official discourse on communication and development is worth noting in a report for the UNESCO entitled "Recent Progress and Its Impact Upon Communication Policy and Development," in which broadcasting, its capabilities, and new viewer demands were elaborately detailed; in significant contrast, the discussion of the impact of such communications capabilities was limited to: "Such an evolution will doubtless have major repercussions on human activities and habits in the aggregate. We may readily imagine the impact this evolution will have upon education and culture, both human activities of primary importance" (Gazin, 1978, p. 15). Elaborating on the possible impact would require responses to questions such as—what might be the nature of the impact, who, specifically, would benefit from this impact, and would such an impact be desirable, uniformly, for all societies?

The generalized rhetoric also included the treatment of the Third World as a single entity, either as a region or continent. The report of the 107th Plenary Meeting stated, "Recognizing that the potential of the field of communication should be applicable to all developing countries . . . allow[ing] all developing countries to have equal access to communication technology in order to develop and operate their own communication system . . . [etc]." Such inclusive address could perform a discursive function of "plac[ing] beyond contestation" the role for communications as instruments in building developed societies. In his analysis of colonialism, Spurr (1993) explained this strategy of "rhetorical appropriation" as "domination by inclusion... rather than confrontation" (p. 32). The reference to equal access in this quotation, one of the major issues in the NWICO debates, does not elaborate on important specifications about access. Consequently, critical questions such as access in which domains, whether access includes information databases or news markets of developed countries, and so on, require answers. For instance, such visions of equality could be based on the communication opportunity structures available to the advocates of the free flow of information and a free marketplace of ideas. Lummis' (1992) interpretation of the concept of equality is helpful here. For him, ". . . equality of opportunity only makes sense in a society organized as a competitive game . . . the rules of the game [are equal] . . .The idea is that the division of society is fair if it takes place under fair rules. Equality of opportunity can thus be seen as a device for legitimizing inequality" (p. 43). Where information and communications have been established as a resource for development and are embedded in transnational capital (as Pavlic & Hamelink, 1985, demonstrated), it problematizes our understanding of equality of access, which implies equality of peoples—something that development promises eventually.

Finally, attempts at cooptation are evident in the instance of satellite communications. The report on the conference held in Addis Ababa in 1981 contains numerous references to the African continent, satellite communication needs for Africa, and the benefits that Africa would experience were it to adopt the recommendations of the conference. There was an assumption that all of Africa had the same or similar needs for communications technology and satellite communications. Treating the continent as a single unit enabled planners and experts to work within the discourse, thus marginalizing the different histories and colonial experiences from the scope of considerations. These are some ways in which the discursive strategy of macrospeak maintained difference in the communication capacities, and deferred the ideal democratic and developed society for developing countries.

In the process of constructing the difference between the First and Third Worlds, the temporal dimension is controlled and contained in this strategy. New communication technologies are introduced to design and shape developing societies, and old ones are deployed toward the same end. The discourse moves away from tradition (the past), which in many ways was necessary to Third World countries. But deferral of development, caused by the inequalities structured by both the materiality and discourse of communication and development, made development a utopian goal. In this enterprise, one useful interpretation of modernity—that of a sense of "belonging to the present," away from the past, but also away from the constantly deferred future—is missing. (Rahnema, 1992, 127). The normative tone implies a continued journey for the Third World toward a predetermined destination in the future. What comes across is the idea that Third World states were expected to defer entertainment or other reasons for communication and cultural expression, and concentrate on information and education for development. Phrases such as application of media for development "should be considered" (107th Plenary Meeting of the UN General Assembly in 1979–1980 imply that broadcast and new media technologies should fundamentally serve the development project, and new technologies in particular are seen as providing the answer to their communication needs.

Commercial interests and imported entertainment notwithstanding, this section of the Report implies that developmental information meets the real needs of people, but pleasure and leisure associated with communication are not addressed. It reminds us of a parallel to economic austerity—a cultural austerity in the media age. Thus, the discursive strategy served to demarcate developed from developing countries, and emphasized the mandatory shift across this divide for the latter. The scope of the generalized statements on communication and development, combined with this normalizing and normativizing tone, tend to make certain definitions of communication and development less questionable and more acceptable as a natural destination for all societies.

Cooptation of Participatory and Expressive Cultural Practices

Certain premodern communication practices and a more community-level democratic idea of participation in development communication projects were absorbed in the larger discourse of communication and development. This involved, primarily, integrating alternative approaches to communication such as participation and also traditional culture in the service of development. In the previous trope, I discussed the detailed descriptions of the traditional media and their uses for development. In this section, I demonstrate the ways in which traditional media were subsumed into the development discourse.

A concept that held special appeal for developing states at this time was the idea of conscientization. Introduced by Brazilian educator and philosopher Paolo Freire, *conscientization* refers to an education practice that stirs critical self-awareness among the underprivileged, of the oppressive structures to which they are subject. The term conscientization was used frequently in the NWICO debates. For example, one document (Verghese, 1978) called for stirring critical self-awareness among peoples of developing regions through central communication planning and development structures. Originally intended by Freire primarily for grassroots literacy and education (or a redefinition of *education*), the concept of conscientization, according to this recommendation, would become a state-planned and directed development activity. As an idea that advocates for consciousness of multiple sources of oppression, if coopted into the dominant discourse of development, it is in danger of losing its edge as a grassroots democratic path to social change.

Among what have been called *expressive cultural traditions* can be located many rural, folk, and other forms of traditional media. Cooptation involved identifying, researching, and documenting these media, with the intention of gauging their potential as instruments to be integrated into the service of development. A UNESCO report discussed earlier in the chapter identified traditional media, their form, structure, and the theoretical entry points that these media have to offer. The report suggested changes in some of the traditional media for absorbing them into the discourse of communication and development:

> . . . all possible approaches and methods should be pressed into service. Live performances are as valid as mass media; mass media must use old formats and evolve new ones; structure must be both preserved and destroyed. (Malik, 1982, p. 86)

This statement illustrates the extent of mobilization of nonmodern mass media for a specified type of social change. It raises some questions: who validates live performances? If folk media retain the folkness (more for setting themselves apart from nonfolk forms than as an essential construct) of their performance because traditionally, they were independent of governmental/development agencies' supervision, can supervision "preserve further . . . the unique ramifications [of] folk media" (Malik, 1982, p. 86)? The document advised retaining uniqueness of form, but coaxing into it a different content, which places the consideration of folk media in society in a dilemma caused by the tensions of the need for economic development and the desire to avoid cultural imperialism.

One UNESCO report on participatory community communications also illustrates attempts to manage alternate communication strategies. I provide a close reading of this text as an instance of the absorption of participatory communications in the broader discourse. The dominant discourse apparent in this document is the initiation and sustenance of a polemic within acceptable bounds. Comparing existing and past projects, the report contrasted the participatory model with the then predominantly nonparticipatory use of communication for development. Here participatory communication involved entrusting development efforts with people and communities, where they would make decisions about communication messages, strategies, content, and other issues that would be appropriate for the community context. In the report, access and participation were embedded in what might be identified theoretically as the tension between direct participatory democracy and representative democracy. The limits of access were defined as "the individual's right to know and right to be heard . . . the logical limitation of communications access" (Berrigan, 1979, p. 9). This text contains some unusual departures from the dominant discourse on communication and development, such as the idea of participation in community communications as a sign of development. But certain statements and ideas urge another reading that is more in keeping with the dominant discourse. The report acknowledged that

> . . . the community media approach does not deny the necessity for the continuation of some or all of these functions [eliciting support for development projects, encouraging family planning programs, etc.] for the media. There is still a need to inform, and to point out the reasons why certain development programmes are being undertaken." (Berrigan, 1979, p. 10)

Explaining the reasons for taking certain development programs is a communication activity here that elicits legitimation for development activities within the bounds of the dominant discourse. It argues for community involvement, people's self-expression, and informed citizenry through grassroots education. Without a doubt, development advocates from all sides of the table would agree that these are critical for the well-being of developing societies, but because of dominant modes of development intervention, this idea of good would conform to dominant definitions of *development*.

Using the example of the Audio Cassette Listening Forum Project (ACLF) launched in rural Tanzania, where the objective was to develop self-reliance in women of the selected villages, the report endorses their optimization of their communications resources and participatory communications activities for identifying needs related to problems of water supply, health care, insufficient nutrition and clothing for children, and migration of young people to urban areas, among other issues. But the author pointed out that they did not engage in needless and even self-defeating political communication, which could potentially entail critique of the existing system. Thus, participatory communication was used to solve a development problem at the community level (no doubt critical to the health and survival of the community), and there were some radical moments—community women's participation in identifying needs, providing solutions, and monitoring activities—but an optimistic belief in the dominant idea of development lingered. Politicization through the media then seems to be reserved for the developed countries, and a practical, action-oriented approach to development was advocated for the Third World, with problem identification and solving being the primary participatory communication activities. Such politicization perhaps would, as with other areas of development, be deferred.

Additionally, this report called for participatory community communication suggesting an assumption that such participation was absent or at least not present to a degree that experts approve. The intervention of state agencies and other powerful institutions to initiate such participatory projects has been interpreted elsewhere as power centers' invasion of "vernacular spaces" (Rahnema, 1992). Rahnema maintained that, through this intervention, participation has become "disembedded from its origins in socio-cultural roots and instead is seen as an activity that would contribute to keeping the economy alive." Further, he pointed out that "it is a tautology to state that . . . . traditional [societies] are participant" (Rahnema, 1992, p. 120). Overlaying organic community communications with participatory communication introduced by the various development institutions could marginalize indigenous community communications' capacity to address non-development issues that are important to the community.

CONCLUSION

From the prior discussion, it is apparent that several documents selected from the 10-year period for this study can be located in the dominant discourse of communication and development. These texts were produced in the decade between 1976-1985 and the NWICO debates constituted the broader context. For example, Document 1 on traditional media in India was published in 1979, the year prior to the official publication of the proposal for a NWICO in the MacBride Report. In the same year that the MacBride Report was published, the document on traditional media in Egypt was also published. Two years later, closer to the impending withdrawal of the United States from the UNESCO, Document 2 on traditional media in India was published.

To a large extent, proponents of new communications technologies and the free flow of communication emphasized the importance of technology. They argued that it was technologically advanced communications that would eventually guarantee democracy -- to adopt Habermas' terms, a global public sphere. To even a staunch defender of modernity and enlightenment (associated closely with Western technological advancement) like Habermas, technology presents problems for democracy. Although his adherence to reason and true enlightenment has come under critical scrutiny (e.g., Poster, 1989), Habermas' argument that the intrusion of the system (here the state, operating through the "steering mechanisms" of money and power—Dews, 1986; in the context of this research, we can add donor agencies and suprastate institutions) into the lifeworld and its consequent limiting of effective communicative action (achievement of understanding through reasoning and argument) need to be noted. In the case of developing regions, scarce economic resources and the threat of transnational corporations' invasion of national sovereignty encouraged state control of communications technology for the purposes of development. Overemphasis on communications technology for achieving development and eventually democracy, then, slows down or perhaps even hinders the moves toward achieving a true global public sphere.

Both tropes—enumeration and macrospeak—produce what Foucault (1972) termed a restrictive economy for conceptualizing a place for communication in developing societies. Different types of texts—in the instance of this project, official, diplomatic, and academic—participated in and (re)produced this restrictive economy: "One shows how the different texts with which one is dealing refer to one another, organize themselves into a single figure, converge with institutions and practices, and carry meanings that may be common to a whole period" (Foucault, 1972, p. 118). From the official texts, we see that ritual "authorized a certain vocabulary" (Foucault, 1978, pp. 17–18), a certain way of conceptualizing communication officially, at the international and national levels, for Third World countries. Such authorizations intensified during the NWICO debates as communication constituted a major agenda in the UNESCO at the time. The tropes in this chapter demonstrate the ways in which discursive paths exerted "control over enunciations" through the official perspective (international, regional, national). Elaboration of details and generalized references to future plans and normative statements limited the range of meanings for communication.

THIRD WORLD ACCOUNTABILITY FOR DEVELOPMENT THROUGH COMMUNICATION

Emerging from the modernization paradigm in the 1950s, and its subsequent adaptation by state governments and international development agencies, development communication became part of the arsenal for fighting rural underdevelopment and was used to introduce rural populations to various ideas and practices, such as new agricultural techniques, the idea of credit, family planning, and other activities characterizing the process of modernization. Typically, various media, including mass and interpersonal, new and old, are employed for the purpose of development in this paradigm. In this chapter, the focus is on the use of development communication in various forms—endogenous, ethnic, exogenous—for modernization through which Third World nations would attain roles as able and competent participants in a new international information order.

Traditional conceptions of development communication, with its mission of contributing toward a certain normativized type of social change in the Third World, demonstrate the discursive strategies of surveillance and invisibility. During the NWICO debates, the rhetoric of communication for development, mainly with reference to new but also at times traditional media, acquired special significance. Following as it did in the wake of the demand for a New International Economic Order (NIEO), much of the rhetoric in the NWICO debates continued to see development in communication as a means to deliver Third World nations to an information-independent status similar to the one that the First World enjoyed. Thus, communication policy for development, and development attitudes and considerations toward building communication infrastructure constituted part of the idea of a new international communication policy and played a key role in the discourse of communication and development.

Like the tropes of enumeration and macrospeak, surveillance and invisibility also critically demonstrate the existing vertical picture of the world, with the countries possessing technologically superior communications capacities and functioning in relatively free market economies occupying the position at the top. The shaping of the discourse of communication and development through enumeration involved detailing; cooptation and normativizing operated through the trope of macrospeak. In the discussion of the tropes in this chapter—surveillance and invisibility—I attempt to discern what emerges as visible within a frame that sets the terms for interpreting communication and development and what tends to remain outside the frame; patterns and reinforcements of visibility and invisibility led to reproduction and naturalization of specific ideas of development through communication. Surveillance and invisibility operated under the broader metaphor of the gaze.

I examine the trope of framing in the discourse of two sets of actors: (a) the critical policymaking groups, media owners, and others from the developed world invested in the advancement of what Escobar (1995) termed the *development project*; and (b) the decision-making élite and other supporters of the mainstream notions of development who were also participants in the dominant discourse of communication and development. These groups, from their vantage points, enframed regions that were perceived to be needy of developmental aid (Escobar, 1995) in various ways. Framing involves the creation of a defined space and placement of objects in it for full visibility and scrutiny and, typically, positioning the creator of the frame outside of it.

In the context of the recent history of communication and development, the agents and institutions involved in defining dominant meanings of the idea of development have worked to make the object of study—the developing regions—transparent. The argument is that full knowledge of such regions is required to be able to aid them along certain paths toward certain definitions of progress. Adopting this strategy of power that shapes the relations between developed and developing nations was not necessarily planned or intentional. However, a particular ideological bent tended to inform language, meanings, institutions, and social practices that constituted the discourse.

Critical works on development in the 1990s increasingly used the metaphor of the gaze in relation to power arrangements in societies, drawing from Foucault's analysis of the principle of the panopticon. Spurr's work on the discourse of colonialism and Escobar's work on power and visibility in development are two notable examples. Conceptually, the architectural features of the panopticon point to a "technology of power." As Spurr (1993) explained, the panopticon "has bearing on any occasion where the superior and invulnerable position of the observer coincides with the role of affirming the political order that makes that position possible" (p. 16; see also Foucault, 1980). The fact that those who are visible to the observer are not in a position to return the gaze is telling of power relations. The panoptic view enables scrutiny of regions and nations within a "field of visibility" (Escobar, 1995, p. 196). Communications media, at times, are used as tools for surveillance, and, at other times, media technologies constitute an index of development. Thus, they become both the object of the gaze as well as the instruments facilitating the gaze.

Two types of power are apparent in the panoptic view. One is the power to survey, and the other is the power to remain invisible. The power and freedom of the gaze, to direct and control, derives from "the *security* of the position from which it [the gaze] is directed" (Spurr, 1993, p. 23; emphasis added). Escobar (1995) proposed, for example, that accountability to donor nations by way of periodic reports, results of studies, and justification/evidence for expenditure of aid bring the Third World under the surveillance of the First World. The trope of invisibility is treated in this chapter as a "deconstruct" of surveillance; those in power survey the communicative spaces of those lower in the international hierarchy while reserving for themselves the privilege of being invisible observers. The ideology of the latest communications technologies as universal requirements, certain attitudes toward the practice of communications (functional, for the purposes of development and/or cultural preservation), and altered relationships with traditional media therefore become natural. At the risk of making an artificial separation, I treat surveillance and invisibility as two inverse tropes serving the same purpose to better understand how each worked to secure the dominant discourse of communication and development.

SURVEILLANCE: OVERSEEING COMMUNICATIONS, OVERSEEING DEVELOPMENT

As a discursive trope, surveillance operates on the premise that "for the observer, sight confers power" (Spurr, 1993, p. 16). Surveillance was manifested in at least three ways during the period of the NWICO debates. First, in official UN reports and articles in policy-oriented journals, satellite communications received considerable attention. Whether for predicting weather for harvests and famine, for detecting natural resources of a country or region, or for any other development activity, satellite communications were highly and consistently recommended for economic progress, but the costs and affordability of this form of communication were not as easily or as frequently visible within the frame (Schiller, 1979). Second, research and surveys of communication resources that Third World countries possessed also brought communication capacities, media use, and other communications activities in the developing regions within the framework of visibility. Finally, evaluations of communications projects that were initiated by intergovernmental bodies like the UNESCO, strong recommendations for periodic progress reports on communication interventions from the aid recipient nations, and the encouragement of self-help (against the extremely limited resources available for investment in the scope and scale of communications projects required for development and the efficient operation of new communication systems) also meant a close scrutiny of developing regions by donors of various kinds of aid. Surveillance is thus a partial explanation for the gaze creating an inequality in the treatment of communication in development.[1] The trope of surveillance allows us to see how communications brings the developing regions into "the field of visibility" (Escobar, 1995, p. 156). The vantage point that the powerful countries occupied, especially with specific technologies such as satellite communication, secured the privilege of surveying. (see Table 3.1).

TABLE 3. SURVEILLANCE

Methods and Activities	Themes
Promoting dependence on and satellite communications technology	Transparency of geographical and natural resources of a country. Strategic pieces of visual communication as data for purchase from advanced nations. Similar close scrutiny of traditional media for strategizing entry points for development messages.
Research to study communications facilities in Third World countries.	Discovering and measuring Third World communication resources. Close scrutiny of communication resources through a lens strongly favoring economics.
Evaluating projects to learn to improve future development projects.	Separation of communications activities from social structure. Evaluation of various efforts such as participatory communication projects and the introduction of media in rural communities.

---

[1] I have adopted Spurr's explanation of the political and cultural economy of colonialism to conceptualize this treatment of communication for development.

Satellite Communication and Transparent Economies

Satellite communication, like most other communication efforts, was expected to help in achieving economic growth in the Third World. Economic growth consistently emerged as a synonym for development. The NWICO debates were initiated with the hope that a new information order would help considerably in achieving a new economic order. Although an overtly economistic discourse as a dominant meaning of development may come as no surprise, through the trope of surveillance it can be seen as a mechanism for keeping in place certain ideas about the role for communication in development. Efforts to make subaltern spaces and their resources visible involve dispelling the subalterns' mysteries. Communication, especially certain communication technologies, allow(s) this exposure in both the visual and other symbolic realms. Discussion on satellite communication and its uses for development rendered Third World economies transparent. For example, for determining the location of natural resources, discussions focused on making geographic terrains in the Third World as completely visible as the technology would allow it. The idea was to use the latest communications technology to accelerate the pace of modernization and development. The metaphor of the satellite technology was extended, at times, into the domain of the traditional media as well, again for making visible all the resources—material and symbolic—for the ultimate economic good of developing regions.

Prominent among the discussions of satellite technology for development were the reports on the UN seminars held in Addis Ababa and Buenos Aires. In the seminar in Addis Ababa, an international group of scientists and development specialists described in detail how satellites would aid in locating natural resources:

> [Satellite communication] will enable us to evaluate the extent, classification, quality, temporal and spatial changes of these [natural] resources. . . . Space platforms and their associated sensors provide us with a unique capability to see and interact with large parts of the earth simultaneously. (United Nations, 1981, Addis Ababa, A/AC.105/290, p. 4)

Chart VI of the document specifies activities such as fact-finding missions, reconnaissance surveys, prefeasibility studies, feasibility studies—military surveillance activities enabled by satellite communications technology. Another chart entitled "Possible Component Sections of a National Remote Sensing Centre" specifies potential areas of the economy where such technology could have a significant impact—for example, agriculture, rural demography, economic geology, forestry, livestock, and water (United Nations, 1981, Addis Ababa, A/AC. 105/290, 4). The usefulness or importance of the technology for a multitude of purposes is not in question. The specification of these communications activities and target areas for surveillance support an ideology that promotes technology and economics as the main constituents of development. Attempts at discursive closure make communications technology imperative if societies were to survive— "In view of . . . benefits . . . and costs and constraints . . . [it is] fundamentally imperative to specify the needs, the requirements of information, and see . . . how remote-sensing can help" (United Nations, 1981, Addis Ababa, A/AC.105/290, p. 4). Nations were thus advised that it was not about inability to afford the investment, but about the risk in ignoring such heavy investments.

Forms of visual communication such as air photos, radar imagery, and satellite pictures produced by various technologies were compared for costs, versatility, accuracy, and extent of information they could yield about resources of a given nation/region/continent. Experts specified levels of visibility—for example, high contrast in satellite images for distinguishing areas with surface water and vegetation (United Nations, 1981, Addis Ababa, A/AC.105/290). "Image enhancement" would "optimiz[e] display" of data for the analyst and would dispel many mysteries surrounding a region's natural resources. Less contrast in pictures would indicate forest types,and geological details such as rocks, and soils, and the least contrast in pictures would indicate other substance classified as biomass. For these types of images, experts advocated skilled interpretation by highly trained personnel as a requirement. Procurement of funds for training personnel did not form part of the agenda in the conferences. According to the experts, the data would be used for managing resources and for protection and conservation purposes. To enable the use of satellite imagery in cartographic surveys and thematic maps, strategic pieces of visual communication on a nation's natural resources that have profound influences on its economy are converted to data; the data then constitute a type of manufactured resource. Because this type of resource production requires large investments, the archiving of such information was taken over by countries able to afford them (mostly those in the category of advanced economies); the knowledge had to be purchased by the rest of the world—in this manner, satellite technology was expected to rescue nations from underdevelopment; this in turn encouraged dependence on the more powerful countries, consequently sustaining unequal international relations.

At the time, two remote-sensing centers were established for agricultural (in Rome) and nonagricultural (in New York) satellite data. These two centers would serve as archives for collecting, storing, and dispensing (selling) data. Thus, full visibility of all resources pertaining to agricultural and other resources would be available to these centers located in the First World. These centers would:

(a) Catalogue, store and interpret remote sensing data, providing facilities for examination by interested parties;
(b) Circulate available information and direct requests by countries for remote sensing data to proper sources;
(c) Provide impartial advice and assistance to projects, Member States, and UN bodies; and
(d) Organize specialized training courses for users, managers and decision-makers. (United Nations, 1981, Addis Ababa, A/AC.105/289, p. 15).

The costs incurred by many Third World countries in purchasing these data have been a thorny issue with client nations; international conferences addressing this issue emphasized the necessity to make the data more accessible to Third World countries by lowering costs. These debates highlight the surveillance nature of the satellite data.

The vocabulary used in relation to modern satellite communication apparatuses in general was also extended to traditional media by the state in developing regions. A more functional role was given to traditional media from the state's perspective. The practitioners of those media and their local communities could assign it one also, but would likely integrate that role differently with expressive traditions. One document described the steps taken by the Indian government to identify, treat, and utilize traditional performing arts for development communication. Part of the process involved experimenting with various traditional media to "understand their nature and communication potential" by clearly identifying all the entry points available for inserting messages determined as necessary for rural populations (Ranganath, 1979, p. 5). Similar examples are available in descriptions of the mass media in Egypt (Hussein, 1980). Entry points for development messages within the folk message were identified (such as the use of religious leaders to ensure source credibility or taking advantage of the mind-altering nature of the messages that could accommodate development messages at a subliminal level). Such exposure also could constitute traditional media as a field of visibility to enable assessment of their functional value for development. Another report on traditional media for development described a chain of activities that would allow the center to use experts to scrutinize, evaluate, and check songs with development themes to be performed to rural populations by select folk media groups (Malik, 1982).

The intention here is not to invalidate the contributions of these documents toward understanding and theorizing indigenous media (these documents provide rare and valuable insights into these lesser known media forms), but to raise a question as to how much of it has contributed to shape the dominant discourse of communication and development. Such analyses of traditional media run the risk of assuming that rural populations do not possess sufficient capacity to survey and commandeer their own cultural expressions for needs they might consider a priority. Examples that have proved otherwise are discussed more elaborately in chapter 5. The urgency implicit in these documents that address the power and potential of traditional media for mainstream development competes with the possibility of other modes of and reasons for existence of indigenous media.

Another strategy to bring rural populations into the field of visibility involved considering the communications satellite as a *change agent*. Satellite communications advocates foresaw dire consequences, such as a return of the dark ages in societies refusing to participate in the technological march forward. One such observation read as follows:

> . . . human progress is largely a one-way gate. We move inexorably onward toward more complex technologies and more complicated economic, social, and political arrangements. The only alternative to continual progress seems to be catastrophic societal failure—the equivalent of a dark age or worse. (Pelton, 1983, p. 78)

An executive of Intelsat at the time of writing this article, the author was a strong advocate of satellite technology. An unequivocal declaration of development within a framework of technology, a unified vision of a global society, and a declared faith in a specific type of communications and communications technology as the inevitable future of mankind work to secure the discourse of communication and development. According to the predictions in this document, the effects on human settlements, work, industry, lifestyles, and entertainment would be unprecedented, where there was potential for the whole world to be transformed into a single modernized and technologized entity (two decades later, we are living that age, at least to an extent, but this has not brought about the corresponding desired or envisioned social change in many parts of the world). Change agents, in the diffusion of innovations theory, are a link between external sources working to induce change in rural communities and the members of these rural communities. Change agents are members of the community they serve. As carriers of development messages and feedback, change agents use their own membership in the community and the knowledge of the community's language to translate the benefits of development to rural populations. In this context, in framing the communication satellite as a change agent, the technology is invested with the ability to conduct almost face-to-face communication, to act as an opinion leader for the community, and possesses an almost human capacity for persuasion. The presence of this change agent amid rural communities facilitates the visibility of the community members and their activities through carefully monitored projects, aided by satellite technology, for the purposes of rural communication development.

Communications Research as Lessons for Development

Research activities in relation to social change are critical to understanding the process and effecting desirable change. This section does not call for elimination of research activities in development communication. Instead, here I contextualize certain forms of communications research that constituted the standard within a certain discourse, and I demonstrate the ways in which this activity has contributed to the maintenance of a dominant discourse of rhetoric and practices that have not been successful in effecting the envisioned large-scale social change. A plethora of research projects and initiatives on communication for development has been generated in the past few decades—the MacBride Commission studies being just one case in point. Evaluations of existing projects' successes and failures, and ways to strengthen future projects, have been offered (e.g., Hornik, 1988; Stevenson, 1988). Knowledge about communications in the developing world—infrastructure, capabilities, potential, target populations—have been discovered, measured, analyzed, and evaluated through types of research emerging from traditions and philosophies at times radically different from those of the developing world. In the process, communications capacities of the developing regions become a field of visibility. The activities of producing such knowledge, setting criteria, and judging accordingly find an analogy in Spurr's explanation of "noncorporeal" power through the use of the panoptic principle (Spurr 1993).

A study on communication indicators of socioeconomic development (UNESCO 1979, CC-79/WS/134) demonstrates this method of bringing communications capacities into a field of visibility. It represents a type of knowledge production that lends itself to control of communications needs for achieving certain types of development. The purpose of the study was to produce a "handbook" for "various types of analysis of existing data" on communications (UNESCO, 1979, CC-79/WS/134, p. 3). The research team questioned the lack of integrated development of all sectors (distribution as well as growth) in the developing regions, which resulted in slow and, at times, reversed growth. The report stated that the future "growth of communication in developing countries must be set within some of these negative trends" (UNESCO, 1979, CC-79/WS/134), a solution that suggests adjusting communications growth within this context rather than addressing the problem at the root. Limitations of the research were noted at the outset with explanations as to why models could not be built and hypotheses could not be tested with the available (inadequate) data. About 103 variables were identified and categorized. Categories included communications, education, transport, urbanization, income distribution, industrialization, technology, growth potential, demography, and an "other variables" category.

Unlike economic indictors that could be standardized for comparative purposes, noneconomic indicators did not lend themselves to comparisons and theoretical bases for causal predictions, and hence they were considered part of the "[overall] poor data base" (UNESCO, 1979, CC-79/WS/134, p. 12). Again drawing an analogy from Spurr, (1993) and reading researcher for writer: "The writer is placed either above or at the center of things, yet apart from them, so that the organization and classification of things take place according to the writer's own system of value" (p. 16). Inadequate databases frustrated the attempts to learn about communications resources available for comparison, improvement, augmentation, and development. Two measures of comparability would have to be established—comparability in relation to the criteria set by the developed countries, and the ability to compare developing countries and regions with each other. Opening up details about communication systems would help with such comparisons.

In the case of socialist countries of the Second and Third Worlds, where "significant growth of communications relative to their GNP" could be discerned, the report stated that the "quality and content of the media" needed to be rated (UNESCO 1979, CC-79/WS/134). Measurements of communication included newspaper circulation per capita, newsprint consumption per capita, telephones per 100,000 population, and radio and TV receivers per 1,000 population. This discrepancy in criteria (quality and content of the media, as against other measurements) perhaps enabled socialist countries to keep out of the frame. Although such data indeed have their uses and contribute to understanding the communications picture of various nations and regions, the underlying universal nature of communication as a developmental resource, and the assumption of comparability of regions within these established categories reflecting a particular history of industrialization in a different part of the world, serve to mainstream certain notions of communication and development. Universal criteria enabled a single broad measurement of communications growth that could be mapped, planned, and monitored. This type of research, when established as the norm, enables control over visibility of communications resources that can subsequently be mobilized for prescribed types of social change.

Another discursive strain involved the integration of communication and economics. That communication is necessary to conduct economic activities, as Babe (1993) argued, is not in question. But the integration of the two in the context of the dominant discourse of development where economic development is necessary to develop communication makes the latter primarily an instrument for economic development. The results of the efforts to scrutinize the role of information in economics are treated as lessons in development. In a workshop on the economics of communication held in 1980, with participants representing the disciplines of economics, information science, communication studies, and development studies, the aim was:

> ... to bring together economists and communication scientists to determine a methodology for integrating communication variables into economic development models. ... Expertise [available] at the workshop would be directed toward formulating market effects, distribution effects, and diffusion effects of communication technology as a variable in economic development, planning, and investment policies. (Jussawalla & Lamberton 1980, p. vii)

A seamless integration of national systems in developing regions merges into a single economic system that could be controlled for achieving a specific type of development:

> It was intended that this activity would ... bring about a better conceptualization and measurement of the communications sector as a macro input for development. Thereby it would assist in framing a general theory for matching the demand and needs for communication as a development agent in national and international systems. (Jussawalla & Lamberton, 1980, p. viii)

This report argues that information is not a commodity, but a resource that should feature as such in national economic planning and policy for development. Complete knowledge and control of resources is a prerequisite for economic planning. This knowledge constitutes a field of visibility in this context when it serves as an agent of modernization and development in the international system.

Accountability in Project Evaluations and Progress Reports

Planned communication interventions, such as the Satellite Instructional Television Experiment in India, the radio project in Kenya, and others allowed the donors of financial aid, expertise, and technology to evaluate projects and monitor the progress of these interventions from their vantage point as donor entities. Often such evaluations ended with recommendations for improvement. The less developed countries were accountable and documented their efforts which were then scrutinized by those qualified to evaluate a project as successful or unsuccessful, and issue recommendations that suggested improvements to set societies in the direction of specific stages of social change.

Integrating community media and community participation in development efforts was an important part of the UNESCO mandates for development communication. A report reviewed and recommended participatory communication and the use of community media (Berrigan, 1979). The Community Media Methodology section of this report contains complete descriptions and evaluations of selected projects. The report included background and context, aims and objectives, methodology, and evaluations. It recognized community empowerment through participation, which involved, in part, allowing the community members to determine their needs. In general, the report recognized the value of participation, enhancement of self-esteem, community pride, and other outcomes as nontangible, indeed nonconventional, indicators of social change. But in the nature of development projects, resources had to be carefully distributed; limitations were placed on what the project could allow; these limitations enframed a field of visibility for project coordinators; and within this enframed space, communications for rural communities are researched and reported.

The Audio Cassette Listening Forums (ACLF) project instituted by the U.S. Agency for International Development (USAID) in Tanzania in 1977 as a participatory communication project for women's development is worth revisiting in the context of this trope.[2] The goals of the project were to help women recognize the importance of their role in society and to take independent and effective decisions on issues related to nutrition and health. Although the goals were to put into practice an alternate mode of communication, the field experiment research design brought the idea into the fold of surveillance.[3] Two villages were chosen to implement the experiment, and two were designated as control villages. The emphasis in this project was on the participatory method, and the groups were created to enable monitoring the method's effectiveness. The village women were responsible for generating media content following a training program. The women in the intervention villages were divided into groups, and group leaders were trained in various aspects of the projects. Interviews with the women and their profile of needs were collected over the period of a few days, and these data were compared with the survey administered by the project coordinators. Audiotapes with group discussions and a performative style of the sociodrama on needs identified by village members were played to other women in the villages, and discussion followed the listening sessions.

Pre- and posttests of the ACLF project revealed mixed results. Changes were an outcome, and participants expressed increased self-awareness and pride through participation. What was not clear from the documented results was the effectiveness of the audiocassettes in stimulating dialogue. Participants of one of the villages belonging to the experimental group were observed to be apathetic to information discussion tapes, but showed more

---

[2]I have discussed one of the projects described in this report as an example.

[3]For a historical and conceptual development and critique of the use of scientific approaches to social modernization and development, particularly in the United States between the 1950s and 1980s, see Tipps (1973). Also see Rahnema (1998) for an excellent and comprehensive review of the various approaches for achieving development.

animated response to problems that were acted out on tape (sociodrama). Because the outcome of the intervention on directly addressing identified needs is not clear, it becomes difficult to sort out the response to the communicative genre (a larger social context for the message) from the response to the development problem and its solution that the genre raised.

The project defined the field of visibility in three ways. First, the villages were selected and designated as the intervention population, which would allow for a close study of the villages. Second, evaluations focused on measuring the success of the project; information on the possible longer term effects of the solutions proposed to problems and solutions emerging through dialogue over the audiotape content do not appear in the report and, hence, remained outside the frame. Third, as the study shows at the end, the projects are not designed to stimulate political awareness or make recipients question the structural nature of inequalities in which they are embedded.

Rahnema (1992) contended that the idea of participation is coopted into the mainstream conceptualization of development. He pointed out that participatory projects allow a "close knowledge of 'field reality' which foreign technicians and government bureaucrats do not have" (Rahnema, 1992, p. 119; see also Huesca, 2003, for a critical analysis of participation with specific reference to communication for development). Hence, the field reality becomes a field of visibility that allows for control of communication activities to lead communities and societies toward the state of modernization.

Another way to incorporate communications in the field of visibility involved separating communications activities from their social structure and fabric, and examining them for their effects on behavioral change as defined by modernization and development. Two case studies involving two development projects in Brazil and Guatemala, respectively, are examples (Contreras, 1980). The Brazilian project (conducted in conjunction with the USAID) specified target audiences of farmers with larger holdings per capita than the Guatemalan project (funded by UNESCO, UNICEF, and USAID). In the projects in both countries, the goal was to employ media to modernize rural areas—specifically, the impact of communication on the opportunity range (circumscribed by factors such as the size of the holding social organization, education, commercial facilities, and communication) for the individual farmer in each of these rural interventions. In the evaluation, the idea that communications alone would not be effective as a modernizing agent because of structural constraints served as the point of departure for this study. To demonstrate this phenomenon, the communication activity was separated, in each case, from other structural factors that were incorporated into the study as constraints. In the process of ensuring heightened visibility of communication activities in this context, the project separates them from the very factors in the opportunity range that make communication as much a social as an economic or development phenomenon.

A result of the study shows that the small farmers in Minas Gerais, Brazil, sought information through the media. The Guatemalan farmers did not demonstrate similar media consumption patterns. The biggest structural constraint attributed to these results was poverty:

> This author's reading of the Guatemalan evidence indicates that there is no reasonable basis for assuming that the factors inhibiting modern behavior are fundamentally cultural resistance or mere ignorance; rather, it may be crudely put as a question of resources. (Contreras, 1980, p. 122)

The project evaluation took into account the contextualized nature of communication in the sense that communication was an integral part of the social structure and the need for it to be treated as such was expressed. It concluded that internal and external tasks needed to be attended to for successful and effective development projects. The external tasks referred to the societal conditions and the internal to communication strategies. In separating communication from societal conditions, the constitutive nature of communication is separated from the (social) picture and made available for close scrutiny in carefully designed development communication projects. This study dissected an intervention; the project was initiated by donor agencies and educational institutions that also represented the development apparatus in this particular context, spelling out the criteria for modernization in their capacities as institutions possessing development expertise.

In this section, the economy (particularly resources) and communications, through inquiry and interventions, served as objects of surveillance for the discourse of communication and development. As Escobar (1995) pointed out, "the role of vision extends far beyond technologies of control to encompass many modern means for the production of the social" (p. 155). Enframing the objects of surveillance and subjecting them to scrutiny, evaluation, and further direction naturalize the goal of producing a social whose meaning emerges from a limited range of definitions for development.

INVISIBILITY: HIDDEN OVERSEERS

Development literature, especially in the 1980s, addresses the subtle position of invisibility. Anthropologists' self-critique of their ethnographic practices as observers, coupled with the idea that developmentalism is a continuation of the earlier European colonialism, help understand this subtlety. Escobar linked the politics of the gaze to knowledge systems. Analyzing the works of Mitchell and Clifford, Escobar (1995) wrote:

> [The] experience as a participant observer was made possible by a curious trick, that of eliminating from the picture the presence of the European observer; in more concrete terms, observing the (colonial) world as object "from a position that is invisible and set apart" (Mitchell, in Escobar). . . . It is reflected in . . . a stand that dictates that the Third World and its peoples exist "out there," . . . to be . . . intervened upon from outside. (pp. 7– 8)

A similar reasoning can be applied to the discourse of communication and development before and during the NWICO debates..

Two interpretations of invisibility inform this trope. First, invisibility can be discerned in terms of the historical roles the economically advanced countries played in the premodern economies of developing regions. Things were set in motion then for definitions of ideal societies and world orders later. Second, it is also possible to read invisibility in terms of the entities to which the developing world is accountable, in reporting progress on development and generally, proving themselves as worthy recipients of aid, and worthy of becoming full-fledged members of a new international information order.

Much of the discourse emerging from suprastate and state agencies at this time suggests that they saw at least three roles for themselves (see Table 3.2). They saw central international organizations such as the UN as *guides* of planned communications in the Third World. Here, development can be interpreted as being guided by invisible hands. The power inherent in the role of the helper or guide is backgrounded by the proponents of development communication (planners, policy decision makers, or the state). The multiple publics, as the targeted recipients of the help, need not necessarily take the guide for granted. This could be due to the short-term nature of interventions, seldom a lasting change for the better (as the critical literature on development demonstrates), or simply that the needs of the publics are not met.

In its role as *parent/adult*, the developed world presents itself as an exemplar to be emulated by the developing world. The guide's tones become admonishing when there is a danger of resistance to their ideas of growth and progress. The developing world's responsibilities for catching up are spelled out by the parent/adult. Another theme in this trope is the developed world's role as occupant of a much mystified position as developed, although the state of development is integrated in to all aspects of social life, this composite position in its (powerful) abstract and ideal sense remains opaque to the majority of the Third World publics. The *mystique*, the dictionary definition of which is the quality of being inexplicable or secret, appears so because of the deferred promise of development that is always to occur at some unspecified future date (Doty, 1996). The deferred society is apparent in the rhetoric about eventual consolidation and establishment of communications in the wake of technological growth and economic development recognizable in terms of the dominant ideology of development.

TABLE 3.2 INVISIBILITY

Methods	Themes
Guiding development communication projects	Authoring development projects. Closely overseeing development projects from design to evaluation.
Playing parent	Inspecting communications capacities and development activities from a superior vantage point.
	Presenting "developed" as a desirable, adult, and mature stage of social evolution.
	Encouraging emulation of the exemplars—original authors of communication and development.
The actual lived state of development as still mysterious, unknown	Hints and suggestions for limitless growth in a developed society as enticement. General, rather than specific, articulations of communication as a constituent of an actual developed society.

Guiding Development Communication Projects

To provide adequate guidance for development, research was needed to produce the required transparency of communications capacities, infrastructure, and lacks in the developing world. But the questions of who researches/guides, who evaluates, and whose recommendations are acted on is not always visible in the discourse. For example, this strategy was visible in project evaluations carried out by development agencies, where the intervention was initiated in such a way as to place beyond question or visibility the main acts introducing the change in the top down approach (see Escobar, 1995, for further development of this idea). (I have excluded Nongovernmental Organizations [NGOs] from this list because their role is another complex terrain that is beyond the scope of this project). Other examples are also discussed later.

In the evaluation of participatory communication projects discussed earlier, the first assumption was that the targeted villages needed to be introduced to the idea of participatory communication (Berrigan, 1979). Although it is acknowledged in the report that the intervention, the media, content, and problems are highly context specific, the need to introduce a carefully designed participatory effort is apparent—the source introducing participatory communication is invisible. Phrases such as "One of the main purposes of the project was to establish . . .," "It is important to recognize that . . .," or "One of the criteria in drawing up the evaluation was that it should directly benefit people . . .," imply this invisibility (Berrigan, 1979, p. 31). These phrases raise questions such as the following—who decided that the project was needed? Who determined the purpose of the project? For whom was it important to recognize the outcome of the project? Who specified the criteria for evaluation? The invisibility of monitors and evaluators of participatory communication implies the construction of difference between the complex of interventionists and the recipients in the developing regions. Similar instances are available in the document entitled "Communication Indicators and Indicators of Socio-Economic Development." The document points out that, "while there have been marked strides in terms of growth, this has not led to integrated development . . ." (UNESCO, 1979, CC-79/WS/134, p. 6). The authority of this evaluation does not come into question; the assumption is that a form of economic development is universal and thus needs no reference.

An analysis of the introduction of radio in Homa Bay, Kenya, demonstrates that the goal of the project was to create a region-specific fit of communications technology and content, and that this goal was successfully met (Mills & Kangwana, 1983). The report begins with a catalog of communications problems facing Kenya, and the country was established as backward on economic and technological criteria. UNESCO-trained personnel (engineers and producers) worked with local engineers to teach them a way of building radio receivers with indigenous material and know-how. Although the idea of a community-specific, low-cost radio may be very viable for whatever purpose the community may choose to put it to use, in this context, the decision to introduce radio in this community emerged from the larger discourse that had established certain modes of communication as facilitators of development. The country became "backward" by certain preestablished standards (in the colonial period, the differences between the colonizers and the colonized, and in the postcolonial period, between the developed First and the developing Third Worlds that came to be naturalized as an ideology kept out of the frame of instability.

Parent Nations as Adult Communicators

The term *parent nations* refers primarily to the cluster of nations that first defined and established the concept of modern development. As "adult communicators" communicating with and setting an example for the developing and yet growing Third World, parent nations acquire the privilege of inspecting and examining the latter, but deny the same privilege beyond a limited extent to the weaker party. We can extend Spurr's analysis from the politics of the gaze in colonial discourse to the discourse of communication and development—the privilege to look at the other but conceal oneself creates an "economy of uneven exchange" with the object of the gaze (Spurr, 1993, p. 13). Critics of colonial relations (such as Doty, 1996) have noted this treatment of developing regions as undeveloped in the human sense also, along with material backwardness. Similarly, the economy of uneven exchange has perpetuated the idea that an ideal developed society that signifies adulthood; in adulthood, people and institutions are facile users of modern media and new communication technologies. Regardless of the combination of elements respective countries should choose as convenient, the overall efforts to blossom into adulthood should be unquestionable, and therefore controlled and orchestrated. Such a parent–child relationship ascertains that like the natural biological growth of humans worldwide, growth of nations would follow a natural, linear temporal path that was inevitable (Doty, 1996, Sachs, 1992). In most cultures, there is a collective social agreement that parents or adults know best. They possess the authority to check, scrutinize, and admonish. Their authority is usually unquestionable, and they are examples that younger entities will have to pattern themselves after. This analogy can be applied to the role of communications in development, as the following analyses demonstrate.

Because modernization and development are symbolized by technology-intensive countries, the parent role is especially apparent in the area of space technology and satellite communications. This role is played out noticeably in regions and continents deemed extremely backward (a problem child). Thus, for it to be appreciated by the African continent, the value of satellite technology needs to be explained:

> Man's venture into space has opened up a new, active and ever expanding frontier, which will, from now on, dominate, to a very large extent our activities here on earth. Space exploration . . . requires adequate investment in training, research and development . . . [now a] practice in all the countries that are developing . . . capability in the area of space science and technology. (United Nations, 1981, Addis Ababa, A/AC.105/289, p. 5)

The report stated that this was the practice elsewhere and it should be adopted here as well, and that the African continent should emulate these countries for its own advancement and for its future. Experts explained that some automated form was required to convert satellite imagery into readable form through digital techniques and manage the volume and interpretation of information/data. But they acknowledged that developing countries were bound to using computer applications developed by the industry, not necessarily with developing regions in mind.

The problem that developing countries raised in the NWICO debates was also largely a question of communications access. Obstacles to gaining access such as the magnitude of financial outlay needed to obtain the required number of computers for reading satellite images, and the technological capabilities required to manufacture customized image processing packages, were not addressed and hence were kept out of the incompetence frame of visibility. When financial resources for satellite technology in the African countries were addressed in the report they appeared as follows:

> For Africa, the question is not whether it can afford, in monetary terms, to participate in space programmes, but can it afford to be inactive in the area of space science and technology mindful of its full impacts and implications on its peoples as well as its group of nations? (United Nations, 1981, Addis Ababa, A/AC.105/289, p. 5)

This statement subtly shifts the burden of catching up on African countries. Another strategy demonstrative of the parent's role is the treatment of Third World countries as a tabula rasa at the time of the First World–Third World encounter (an old critique, but useful for illustrating the trope under discussion). As with children, skills need to be taught socialization and survival in the larger world order. The second satellite communications conference in Buenos Aires called for "rapid development of effective mass communications . . . for 'practical instruction of village inhabitants'" (United Nations, 1981, Buenos Aires, A/AC.105/290, p. 14). Premodern knowledge did not enter the picture. Practical instruction could potentially involve marginalizing or coopting previous knowledges and writing fresh practices through instruction on modern media technologies. The role of the adult that the proponents of modernization and development assume allows them to hold the Third World accountable for their actions—how large-outlay communications capacities are utilized, how traditional media and other mass media are utilized in the service of development, whether these countries demonstrate growth in the annual measures of media development—radios per 1,000 population, newspapers per 100 circulation, and so on—continue to be indicators of development.

Communications and the Mystique of the Deferred Society

In most of the texts analyzed in this and the previous chapters, construction of the discourse focused on the means to development through communication, and the end—an evolved set of political, economic, social, and cultural arrangements characteristic of a developed society. This contributed, in part, to the mystique of development, which is still a constantly deferred end point for the developing world (Doty, 1996); the perceived attractiveness of that end state also added to the mystique. It is apparent in hints at unlimited progress and, most of all, relief for the rural populations from their plight. Phrases like "Man's venture into space has opened up new, active and ever-expanding frontiers" suggest infinite advancement toward a relatively unknown end, and the idea suggests that progress breeds further progress (United Nations, 1981, Addis Ababa, A/AC.105/289, p. 9). Similarly, in the document describing a futuristic global telenet, the telecity hints at eventually absorbing developing societies also in the network. While these ideas were optimistic, they did not address economic differentials in the power to invest in satellites. For national policymakers in developing regions, a global telecity on a large scale benefiting their populations would offer, at best, an imaginary destination.

Official reports on meetings also contain instances of the attention focused, on various occasions, on the means to development—that is, on how communications would work toward achieving development, but not on how communications would constitute a developed society beyond the existence and a high degree of dependence on the media. There is the assumption that the final destination would resemble the developed world, an inaccessible vision to the majority of the rural populations who comprise the bulk of the Third World. This assumption legitimized the need for communications technology and other sea changes in backward regions. The possibility of alternate modes of communication within alternate modes of living was marginalized in this discourse.

In the modernization era, the Third World was expected to reach the state of development with the help of the modern mass media. As the discourse unfolded over time, and developing regions did not see expected changes, variations of modernization made it more difficult to articulate the destination to which the means (communication, the media) would take the less developed countries. However, the basic idea of a globally orchestrated move forward and certain patterns in the assemblage of social institutions and functions underlie these variations of modernization.

As a mode of establishing the dominant discourse, invisibility served to "[embed] the universals in the discourse" of communication and development (Escobar, 1995, p. 160). The authors of the idea of communications for modernization often kept themselves out of the frame, but retained full view the objects of their gaze (Third World communications), This vantage point of being able to "[see] everything from nowhere" (Haraway; cited in Escobar, 1995, p. 155) works to consolidate the ideology of a dominant mode of conceptualizing communications for social change.

CONCLUSION

Escobar (1995) argues that to figure in the discourse is a form of visibility. He established the market economy, international rural development, and women in development as crucial objects in fields of visibility, and he demonstrated the ways in which the peasant population and Third World women are brought into this field. Through the tropes of surveillance and invisibility, we can see that development communication and all that it entails, such as research and intervention projects, seminars, and studies constituted a field of visibility for the discourse of communication and development. Both tropes—surveillance and invisibility—worked together to create a global space where the dependent status of developing nations was established and the role of the economically powerful countries as the providers and mature adults was also taken for granted.

In this chapter, we have seen how viewing but remaining hidden worked in concert to reproduce the ideology of communication and development during the NWICO debates. There are links to the tropes enumeration and macrospeak discussed in the previous chapter. Enumeration aids surveillance. Macrospeak, through its elisions, works to obscure power relations and hence aids in keeping the centers of power invisible. Third World economies, resources, communicative capacities, and plans were rendered transparent; this transparency served as a form of accountability for the aid and encouragement given by the West and enabled by technologies, ideology, and experiences that were products of particular histories. In subjecting the developing regions to these four discursive strategies discussed thus far, the West emerged as the "primary referent" (Mohanty 1991, p. 55). The result has been inequality of such magnitude that, as Narula and Pearce point out, it "stifles the imagination of alternative social orders" (Narula and Pearce 1986, p. xviii). Yet discourse constantly seeks to address its "other(s)." In the following chapter, I read attempts at a conscious self-constitution of developing nations in their efforts to restructure for themselves, on their terms, the world order. In the same vein, they also attempted to articulate an understanding of communication that contests and negotiates the functional, instrumental one apparent in the dominant discourse where its value derived from its contribution to development. In the process of discussing the tropes of negotiation and resistance, some alternative possibilities emerge.

NEGOTIATING READINGS OF "WORLD ORDER" AND "COMMUNICATION"

At the present time, it would be easy, but incorrect, to dismiss the UNESCO debates for democracy (and development) as a failure. Nor can we claim that a discursive closure has ended the reality of the need to achieve both global development (interpreted variously) and communicative democracy today. For example, some of the recommendations that emerged from the NWICO debates have been put into practice, such as the creation of regional news agencies intended to ensure greater parity in news flow and fair media representations, even if they struggle to survive because of financial reasons (Boyd-Barrett & Thussu, 1993). Nations and communities are adopting technologies or negotiating the use of these technologies apparent in ways that challenge the fear of globalizing information technology in the NWICO debates. Another serious effort at resisting discursive closure of the NWICO idea is apparent in the MacBride Round Tables held annually for at least a decade or so (1989–1998) after the original debates petered out in 1985. The MacBride Round Tables continue to keep the NWICO arguments and ideas on the research, practice, and policy agenda and also recommend action for adapting to changing media and social conditions and contexts (Vincent & Traber, 1999). In this chapter, I examine the discursive contestation to and negotiations of the dominant meanings of the term *development* as they emerged from the NWICO debates.

Proponents of the NWICO articulated understandings of communication and development that were different from the dominant discourse. They employed the terms, tools, and technologies of the dominant discourse, but inflected them with their experiences, especially in the instance of the Third World, as historically colonized, underdeveloped peoples. Developing countries and activists and scholars from developed world wrestled with ideas of another development and, in the process, with articulating another world order and another understanding of communication. This other encompasses a range of alternate ways of understanding the notions of development and also communication. Such contestations and negotiations occurred in intergovernmental, professional, and scholarly arenas. The environs and context within which these constituencies function, in a sense, impose limits on the resistance they can offer. Governments debated these issues for their respective nations as a whole; representatives of Third World states in UN and other international fora could be perceived as forming a second tier of domination, particularly from the standpoint of the vast majority of people who are not members of any intergovernmental forum and therefore, at the time, did not have a voice in these debates. At times their officials' articulations of resistance could be understood to re-create the dominant discourse at another level, with similar patterns of domination encoded in their solutions and alternatives to developmental problems. As members of the ruling class, state officials possess the power to create and shape communication policies for the rural majority, particularly in developing regions where the state, to varying degrees, controls production (Randall & Theobald, 1985). From the perspective of the subjects of such policy (the public), it is possible that no real gain is seen or, as some development projects have shown, experienced. Nevertheless, at the diplomatic and inter-state circles and other domains, such as the professional and academic, the intention to consciously refrain from subscribing directly to mainstream communication and development ideas helped articulate alternatives to the existing orders. Some of these alternate approaches have materialized at least in part in recent years. To effect these articulations, proponents of the NWICO worked to formulate other understandings of world order and communication, variously more proximate to or distanced from the legible metaphors of the dominant discourse. In the process, they attempted to negotiate meanings of communication, world order, development, and self-determination within this language.

Redefinitions often involve challenging enduring significations of terms from alternate perspectives. In the context of the NWICO, admittedly redefinitions among the developing nations have not effected a reversal of fortunes in the existing world order. These redefinitions are symbolic of the aspirations of the Third World governments for the material conditions of the international society that would be different from (and not necessarily the inverse of) those present at the time of the debates. The dominant discourse counters resistance through various mechanisms such as cooptation (e.g., appropriating the idea of participatory communication in specific ways for development projects), omission, or other, more overt means such as the withdrawal of the United States from the UNESCO in 1984. Counterimaginings of a world order or communication constituted resistance to the extent that they challenged the existing meanings of these terms on which decisions in the international arena were made and acted upon. An opening for actual change is available, beginning with a change in a social understanding of world order and communication. As Hall (1985) reminded us,

> The signification of events is part of what has to be struggled over, for it is the means by which collective social understandings are created—and thus the means by which consent for particular outcomes can be effectively mobilized. (p. 36)

In effect, a possibility exists for social change guided by new definitions.

The concept of ambiguity informs this chapter. Ambiguity calls to attention that a definite and permanent closure of a discourse, or a "stable and unified structure," does not occur (Laclau & Mouffe, 1985; Spurr, 1993). That the dominant discourse works constantly to secure its position indicates that resistance, the "other(s)" of the discourse, is (are) ever present in the discursive environment. As Spurr (1993) pointed out in the context of colonialism,

> Just as law establishes itself by defining the outlawed, so the very nature of discourse as a framework involves principles of limitation and exclusion and therefore creates the possibility for alternative ways of speaking. (pp. 185–186)

Any discursive closure would at best be arbitrary, and this arbitrariness renders the dominant discourse vulnerable to multiple contestations (Hall, 1980). Thus, ambiguity necessarily entails more than one set of meanings attributable to concepts, ideas, institutions, and practices. The field of possible meanings, understandings, and practices then becomes a far less tidy affair, and the idea of resistance as a monolithic counterdiscourse becomes impossible. The challenge to meanings of communication and development that have emerged as dominant, and negotiations with such meanings involves a composite of positions and articulations. Contextualizing the struggle over meanings, practices, and institutions would aid in drawing reasonable boundaries within which the dialogic nature of the discourse can be observed.

Two themes of contestation and negotiation emerged from a reading of the selected texts for the study. The first attempted to envision a world order that differed from the existing center-periphery construction. Drawing on Giddens' concept of structuration, I have called this theme *restructuration*. In theorizing the phenomenon of globalization, Featherstone (1990) interpreted structuration as a concept where "nation-states are not seen to simply interact but to constitute a world" (p. 5). Further, "render[ing] the world into a singular place—through different historical trajectories" (Featherstone, 1990, p. 6) also finds a place in envisioning another world order. The existing world order was a result of the historical experiences of conquest and colonization. The change that some NWICO proponents advocated involved imagining a world that would emerge from the collective modern communications experiences of countries sharing a history of colonial oppression (particularly obvious in the case of the Nonaligned Movement [NAM] member countries). I read their articulation of another world order as a "historical possibility," as resistance.

A second theme addresses a different understanding of communication and, consequently, communication and development. Efforts to separate communication from its almost purely instrumental role in traditional development discourse are apparent in some texts. Also, efforts to define communication as a more holistic and social rather than technological or economic activity are also discernible. In these tentative definitions, communication is abstracted from mass and interpersonal interpretations for purely development purposes and examined as more of a noneconomic activity. I have called this theme *retheorizing* to describe efforts at constructing another understanding of communication in communication and development. Efforts to restructurate and re-theorize occurred at a time when many international communication research projects were commissioned between 1976 and 1985 at conferences, seminars, and in reports intended to help propose a new international communication order. Conceptualizing a different world and other understandings of communication was inextricably bound up with the agenda of the NWICO debates.

The NAM constituted a notable platform of resistance to the existing world communication order. The NAM consisted of a consortium of countries from the developing world, mainly from Asia, Africa, and Latin America, with the exception of three European member nations. Most of these countries were former colonies interested in effecting changed conditions for themselves from about 1927; Singham and Hune (1987) observed that the movement was "the only international social movement in world politics" (p. 186) advocating peace, disarmament, and self-determination for formerly colonized peoples. Although the NAM's history has been traced back to 1927, new articulations of problems concerning member countries came about during and as a result of the "confrontation between socialism and capitalism" during the cold war years (Singham & Hune, 1987, p. 187). As a result of this confrontation, the movement created for itself a third position in international politics.[1] The nonaligned group of countries consti-

---

[1] For an extended and sympathetic treatment of the NAM and its activities at the various summit conferences, its agenda that has been reshaped by changes in international politics over time, and the juxtaposition of this transforming agenda against the continuity of the NAM's underlying purpose, see Singham and Hune (1987).

tuted an important secondary international arena where newly independent, fledgling nations beginning to engage with international agencies such as the UNO and its branches could express their needs in collective voices. In 1983, a NAM media conference was arranged in New Delhi under the name of NAMEDIA; efforts at both restructuration and retheorizing were apparent in reports from this conference.

RESTRUCTURATION: RECONFIGURING THE CENTER AND PERIPHERY

The concept of restructuration is derived from Giddens' work on structuration. Giddens provides the foundations for understanding structuration as a way to think about the world as "one place" and reproduce this world as one unit (Featherstone, 1990; Lash & Urry, 1994; Waters, 1995).[2] His development of structuration involves the interplay of both structure (the social system and institutions) and agency (the individual or the subject) as simultaneously deriving from and constituting society. Political sense can be made of structuration because

---

[2]This is especially apparent in sociological theories of globalization. In this instance, I refer to two works by Lash and Urry (1994) and Waters (1995). Although both works provide a fairly comprehensive and insightful understanding of the phenomenon of globalization, the authors acknowledged only briefly that globalization is a condition stemming primarily from transnational capitalism and focused instead on the rapid and chaotic movements of images, cultures, practices, goods and people across national borders in the context of high globalizations in the late 20th century. Earlier, a politicized reading of such movements in the communications field yielded arguments about media and cultural imperialism (Lee, 1980; Tunstall, 1977). Active audience theories in communication have since challenged these charges; the conceptual notion of resistance and refashioning that is also apparent in these theories can be applied to receivers and subjects of the discourse of communication and development. Lash and Urry provided case studies of the United States, Britain, Germany, and Japan on the premise that globalization is generated in centers of a transnational cultural economy. That the participation of developing countries in this type of economy is constitutive and dialectic is not addressed in their analysis of the late 20th-century culture industries and industrial culture. Robertson went to the root of the issue by using *meaning*, or the world as a "central hermeneutic," as his point of departure, whereby we are allowed to read the Third World also as a sizeable constituent of this hermeneutic. Hence, its explanatory value for the study. For an excellent analysis of the transnational movement of culture and suggestions for theorizing a global cultural landscape, see Appadurai (1993).

> processes of structuration involve an interplay of meanings, norms and power. These three concepts . . . are logically implicated in both the notion of intentional action and that of structure. Every cognitive and moral order is at the same time a system of power, involving a "horizon of legitimacy." (Giddens, 1993, p. 169)

Robertson (1990) offered an interpretive bridge between structuration and the idea of a world order in his adaptation of Giddens' structuration to formulate a theory of globalization. Although Giddens' concept of structuration underwrites this trope, I draw mainly from Robertson's adaptation of structuration and his subsequent theory of globalization to provide an explanatory framework for the trope of restructuration. Robertson obtained analytical purchase from the concept of structuration by making it "directly relevant to the world in which we live" p. 20). He related the phenomenon of globalization to modernity and modernization. His contention was that, although the idea of globalization has existed since early intercontinental travel, "there can be no denying that the world is much more singular than it was as recently as, say, the 1950s" (Robertson, 1990, p. 20; also Germain, 2000). The world as a singular space fundamentally informed the idea of communication and development, with all its ramifications and implications spelled out in the debates around a new world information and communication order.

Unlike some other sociological theories of globalization, Robertson's understanding of world politics hinges on connections to historical precedents, mainly the universalization of the idea of civilization. Robertson, (1990) said,

> I argue that what is often called world politics has in the twentieth century hinged considerably upon the issue of the response to modernity, aspects of which were politically and internationally thematized as the standards of "civilization . . ." during the late-nineteenth and early twentieth centuries in particular reference to the inclusion of non-European (mainly Asian) societies in Eurocentric "international society . . ." (Robertson, 1990, p. 17)

According to him, globalization is a continuous process, not a static world system where a vertical structure once made cannot be unmade. Neither is it integrated in what Robertson termed a "naïve functionalist mode." Globalization is about understanding "how the global system has been and continues to be *made*" (Robertson, 1990, pp. 17-18; italics original). This, coupled with the opening Giddens offers for the politicization of the concept of structuration itself ("interplay of meanings, norms and power"), makes the concept productive for reading a reconfiguration of the world system from a Third World perspective in the NWICO debates. Although many Third World actors were interpellated by numerous institutions, they in turn, as active agents, attempted to redefine a world order defined by these institutions.

During the NWICO debates, restructuration was envisioned mainly through the modern mass media, which were used to differentiate another world order from the existing one. Here, I read three ways of what I have tentatively termed here as re-structuration (see Table 4.1). First, a remapping of the center was apparent, where the locus shifted to a collective of international communities hitherto referred to as the *periphery*. Second, the idea of global interdependence posited by opponents to the NWICO was appropriated by the supporters of the NWICO; the interdependence was termed *collective self-reliance* and again shifted to groups of Third World countries (the NAM being one such group). Finally, ideas aimed at dissolving boundaries between center and periphery were apparent in suggestions for specific journalistic practices and reconfiguring a world without a center/periphery arrangement. I present the first two points—remapping the center and redefining interdependence—in the first section of the trope of restructuration because they are both conceptually oriented. I present the specific suggestions for changes through journalistic practice in a separate section.

TABLE 4.1. RE-STRUCTURATION

Method	Themes
Remapping the center	Redirecting information flow through regional news agencies
	Using satellite technology and computers for regional exchange and and the construction of a collective self.
Proposals of collective self-reliance for the Third World	Collective self-reliance for media materials, following the information exchange model.
	Consciously differentiating between collective (NAM) self and the larger world order.
Journalistic interventions	Demands for more fair and non-ethnocentric media representations of the Third World.
	Demands to share in editorial decision-making at the stage of gate-keeping in First World news desks covering international news.
	A bid for training of journalists from the South, in the South.

Constructing Another Center

Attempts at restructuration were evident in the resolution of a nonaligned countries' meeting at Lima in 1975, endorsed by the participants of the Dag Hammarskjöld Third World Journalists' Seminar. Participants included journalists from Algeria, Pakistan, Peru, Mexico, and India, among other developing countries. In this forum, practitioners' voices were heard on historical and developmental issues. The Hammarskjöld Seminar report included the need for governments to "... cooperate in the reorganization of communication channels still dependent on or which constitute a colonial inheritance and obstruct direct and rapid communication among non-aligned countries" (Development Dialogue, 1981, p. 116).

Following this statement on disengaging from the residual oppression of the old colonial system, the resolution called for exchange of information on national achievements among the nonaligned countries. In efforts to counter the hegemonic flow of agency news produced by the Big Four (the United Press International [UPI] and Associated Press [AP] of the United States, Reuters of Britain, and Agence France-Presse [AFP] of France), a series of news exchange mechanisms (NEMS) was instituted through regional news agencies such as the Caribbean News Agency (CANA), the Nonaligned News Pool (NANAP), the Pan-African News Agency (PANA), and DepthNews, which would cover in-depth development, economics, and population issues (Boyd-Barrett, 1980; Boyd-Barrett & Thussu, 1993; Gauhar, 1981; Lent & Vilanilam, 1979).[3] A post–cold war report on the performance of Third World NEMs by Boyd-Barrett and Thussu (1993) documented a less efficient system than its participants had intended, due to the reality of continued structural imbalances between the less developed countries and the advanced nations. But the regional NEMs were effective in that they supplemented (rather than substituted) news from the Big Four, expanding the horizontal reach of Third World news exchange.[4] The continued presence and functioning of these Second World/Third World agencies

---

[3]Gauhar (1981) used the term The Big Four, commonly referred to in literature on news agencies that preceded the NWICO debates. For a post-cold war evaluation of the status, performance, and effectiveness of Third World news agencies, see Boyd-Barrett and Thussu (1993). For an explanation of the origins and development of DepthNews, see Lent and Vilanilam (1979).

[4]Although this project deals with the aspect of development in the NWICO debates rather than news flow questions that are frequently the focus of studies related to the NWICO, a note needs to be made in the context of restructuration. Boyd-Barret and Thussu titled their work "Contra Flow in Global News," which suggests a counterflow from the rest of the world to the First or, in a more benign interpretation, a reciprocity of agency news exchange between the First World and others. It is difficult to imagine this actually occurring because The Big Four, in their ensconced dominant

addressed at least partially the need for such information from many developing nations. Taken together, the call for a disengagement from a colonial communication system and a reorientation of news flow among nonaligned and other Third World countries suggest a restructuration of the world from a Third World perspective.

A reading of the NAMEDIA conference reports also demonstrates the Third World countries' visions of a restructurated world. In his presentation at the conference, Parthasarathi (1983) strongly advocated collective self-reliance for media materials among the nonaligned countries. Parthasarathi suggested extending the concept of news exchange mechanisms to include collective television and film banks. Collective self-reliance among the NAM members differentiates itself from the idea of the larger global interdependence, the collective self is differentiated from an other, the existing larger world order. Further, ". . . equally important is our collective responsibility that the means and facilities for the smooth and balanced flow of information are created and made available to all parts of the world" (Parthasarathi, 1983, p. 56). Here, Parthasarathi emphasized the NAM group of nations' role in redirecting international information flow, with no specific mention of the First World, and conferred on the NAM group of countries the task of ensuring a worldwide balanced flow. It is interesting to note that the word *free* associated with communication flows (as in free and balanced), apparent in many reports is absent here.

Participants of the Hammarskjöld Seminar of 1981 defined *another development* (italicized whenever used in the seminar report) as follows: "The satisfaction of human needs and the creation of a habitable environment on the basis of self-reliance and harmony with the values and aspirations of each society." The seminar participants believed that this could be achieved by fundamentally changing the world system (creating a new world system) through the establishment of the New International Economic Order (NIEO). Although this may echo the dominant discourse in that the meaning of development was derived primarily from economics, it was an instance where the journalists articulated a new world system (Statement by the Participants, 1976; cited in Development Dialogue, 1981) in the process. They expected decommodification of information and a "valid interplay of different cultures and national realities" to be the outcome of a NIEO.

These journalists advocated a collective self-reliance for Third World countries to further and promote the establishment of the NIEO and the goal of another development. If the Third World were to act out a restruc-

---

position, may not have the need for yet other sources of news, especially ones that are linked with governmental control (regional NEMs). What Boyd-Barrett and Thussu demonstrated are the more divergent flows of news following the institution of the NEMs.

turation in terms of both economics and information/communication, that order would likely prioritize the concerns of Third World regions from their respective perspectives that may differ significantly from the priorities drawn from the existing world order. This might result in a fragmenting of the world system hitherto unified by structural imbalances.

Galtung offered an interesting prediction of what a restructurated world might look like should the then-existing economic trends continue. He saw the rise of the "Japan–China–Southeast Asia triangle" as a sign of the slow but sure shift of a center to what was until then one of the peripheral regions. As a result of this shift,

> ... the expression "Third World" [becomes] ... a misnomer. It certainly applies to some of the Third World countries more than the others, and opens for a recolonization of most of the Third World, but this time from the world's Southeast rather than its Northwest. (Galtung, 1982, p. 136)

Parallel to the economic changes, Galtung saw changes in communication also taking place. At the time of publication of this article in 1982, and also as Galtung pointed out, as many Third World countries had increased media coverage of their respective nations and had noticeably decreased coverage of the First World, a gradual change had occurred over some decades. Additionally, Galtung also predicted that the First World would have to increase media coverage of developing regions in the terms specified by the latter (in a sense some influence exerted by the developing regions over the gatekeeping process for news about them in the media of advanced nations). In these ways, the vertical nature of the world order would shift to a temporary equilibrium while in transition, following which a new hegemony would be established (Lent 1995).[5] This vision of restructuration implies another world order, with a change in the power center.

Restructuration was also envisioned from the standpoint of high technology. The then-chief consultant for the Planning Commission of India believed that it was possible for the NAM group to configure high technology "to help us move in the direction we want" (Pal, 1983, p. 71). The use of the collective pronoun *we* in the statement, "If we in the South are serious about our concerns, there do exist ... appropriate high-technology ways ... [etc.]," brackets out the larger world order and deals with the NAM group of countries as the central entity of interest. Pal also suggested that funding come from pooled NAM countries' resources. Costs could be con-

---

[5]Lent demonstrated a new hegemonic order at work among the Third World countries, where Indian film and video (one among many examples) have invaded and established a firm hold over the Arab region's mass entertainment market.

trolled or scaled down by benefiting from the expanding accessibility of communications technology in two ways: (a) by taking advantage of the flexibility of modern communication technology to conform to needs of countries of the South, and (b), by profiting from the decreasing costs of certain parts like microprocessors in this new low-cost configuration process. By configuring technology very specifically for the needs of the South, Pal held that it would be possible to avoid duplication of the "outward technological manifestations of the North" and the potential for "real priorities from getting lost in a fruitless attempt" to imitate the North (Pal, 1983, p. 67). He envisaged an independent South, with communication technologies tailored to the needs of the countries of the South.

A major theme of the satellite technology conference reports from Buenos Aires and Addis Ababa was the need for the developing world, to use Pal's (1983) words from another context, to set "goals and priorities...to suit what the technology itself prefers to do well" (p. 67). He saw the power invested in the technology as misplaced, and he urged the NAM countries to take advantage of the flexibility offered by communication technology to disengage from this technological determinism. By taking the conceptualization of communications technology out of its original historical context and placing it in a different one, Pal shifted the center of gravity from the North to the South. For example, he suggested that satellite-based networks were more of a priority in many Third World countries than a full-scale telephone service system. Thus, what the North saw as very basic in a telephone system need not necessarily be so in the context of the South. Prioritizing satellite networks over telephones "suggest[s] a reversal of the historical trend in development or telecommunication" (Pal, 1983, p. 69).

Restructuration Through Journalistic Interventions

The Dag Hammarskjöld Report of 1975 states that the Hammarskjöld Journalists' Seminar participants supported the idea of giving the international order a human dimension. This dimension included the recognition of cultural, political, social, and economic diversity (in that order) and its implications for communication were that:

> The image of the Other should reach each of us, stripped of the prevailing ethnocentric prejudices, which are the characteristic feature of most of the messages currently transmitted. (Development Dialogue, 1981, p. 116)

In this statement, presumably, the seminar participants were referring to the major world media's negative portrayal of most Third World countries in the media in developed nations. The participants understood the need for self-reliance for information and communication among the developing countries, and more comprehensive and fair representations of these countries as effecting horizontal links among themselves. If these needs were addressed by the NEMs, dependence on the Big Four would decrease, and this in turn would contribute to eventually leveling the asymmetries apparent in news flow and representations of nations in the news media.

Another articulation of restructuration is apparent in changes suggested in practical journalistic relations between the North and South (Gauhar, 1981). This change was perceived to have the potential to reconfigure the existing vertical arrangement between information senders and receivers in the international media systems to one of a more equitable structure. The NWICO discourse emphasized the necessity for such reconfigurations, but Gauhar provided specific suggestions for achieving them from a journalistic standpoint. In an evaluation of international communication following his case study on the development of the press in Pakistan, Gauhar cited the (rare) example of one of Britain's leading newspapers, *The Guardian*'s, special section entitled the "Third World Review," introduced in 1978. In an arrangement between the editors of *The Guardian* and editors of various Third World newspapers, the latter could write for the "Review" about their respective countries from a participant-observer's standpoint, provided they agreed to *The Guardian*'s editorial control. This section of the newspaper provided a space for Third World voices. The praise and Press Award for this unusual and bold enterprise notwithstanding, Gauhar saw this move only as a beginning; he expected this move to serve as "a model for editorial partnership between the North and the South" (Gauhar, 1981, p. 176). This would ultimately involve shared decision making between the editors of the First and Third World media. Gauhar suggested a similar arrangement for news agency collaboration between the developed and developing countries, by region and topic, involving shared decision-making. Shared decision making at the stage of gate-keeping changes the journalistic power balance between the developed and developing countries and demonstrates a reconfigured world order at the press/media power centers, leading to potentially blurred lines between the center and periphery in a critical journalistic practice.

A report discussed in a previous chapter bears repetition in conjunction with Gauhar's article. At the international conference held in the Federal Republic of Germany in 1978, under the title "Toward a New World Information Order: Consequences for Development Policy," participants suggested that editors of First World media participate in training sessions to help them recognize and present news about progress and social change in developing regions as newsworthy and legitimate information (Bielenstein, 1978). It was also suggested that Third World editors and reporters be trained to understand and write for the Western media markets. The emphasis was on linear exchanges between First and Third World media practitioners; these exchanges would take place within the bounds of existing (conventional) journalistic practice, but did not involve the more important practice of shared editorial decision making that could change journalistic relations between the center and periphery.

Gauhar also observed that training programs in the Third World required the leadership of professional instructors from the South, and that bringing journalists from the South to training centers in the North was both wasteful and "in many ways injurious" (Gauhar, 1981, p. 177). This statement invites at least three readings: that training by professionals from the South would produced a journalism that was noninjurious to their interests, that trainers of the South are more sensitive to development news, and that the attitude toward development news as a process rather than an event needed to be encoded in the training.[6] Confining most major training of journalists to the South might, to a certain extent, check the reproduction of asymmetries that are liable to occur more frequently because of internalizing the professional ideology of the North. Journalists in the South were more sensitized to the need for development news by their respective governments (Golding, 1978).

In this section, I demonstrated the restructuring of the world order from perspectives in developing regions and generally from the supporters of the NWICO. In the case of the NAMEDIA, an openly political collective brought to the table ideas, suggestions, and examples and, in the process, structured a different world order, where the North was present in that it was constantly invoked but did not possess the center of gravity that it enjoyed (and still enjoys) in the conventional world order. As with imagining a different world, imagining a different set of meanings for communication was also apparent in these debates. In the next trope, retheorizing, I read what might be considered another understanding of communication.

RETHEORIZING: ANOTHER UNDERSTANDING OF COMMUNICATION?

In attempting to retheorize communication, my intention is not to look for an overarching framework that would constitute a new theory (or theories)

---

[6]Interestingly, Verghese (1979), in his UNESCO document supporting the dominant idea of communication and development, suggested looking at development news as a process rather than as an event. This is an instance where strands of dominant as well as alternate discourses can be read in one text.

of communication in the sense of theory building. Instead, I read different understandings and articulations of certain concepts, notions, and ideas that have been theorized extensively in Western communication scholarship. These include public opinion, information, communication, and alternative communications or media. Similar ideas and communication theories exist and have been debated extensively in Western scholarly circles. But most of the texts analyzed for this trope make their case for another understanding of communication within the context of the discourse of communication and development in the NWICO debates. One article draws on works by McLuhan and Innis. But most of the other texts analyzed in this section were authored specifically within the context of the NWICO debates, and it is against the backdrop of the traditional conceptions of communication for development that I read another understanding of the human and social practice of communication.

Three sets of ideas were read for the trope of retheorizing (also see Table 4.2). First, different interpretations of the concepts of public opinion and information were apparent. In the discourse of communication and development, often public opinion was seen as a lack or characterized as inadequate in developing societies. From the time of the prescriptions advocated by Lerner (1956), modern mass media as well as traditional media were treated as vehicles for the creation and maintenance of public opinion. In this section, we see how Third World journalists and various scholars redefined the

TABLE 4.2. RETHEORIZING

Methods	Themes
Expanding public opinion and information	Expanding the idea of an informed electorate to an informed public about developing countries, in the First World.
	Emphasizing information with comment and ideological conviction as a more meaningful form of communication.
Returning sociality to communication	Arguing that prevailing notions of the "communications revolution" is a change in tools, not in message.
	Emphasizing host culture context as the major determinant of successful adaptation of communication technology.
Alternative communications	Bringing to the foreground marginalized populations and the popular in resisting dominant media/fashioning new forms of indigenous information and communication activities.

concept of public opinion in the context of the NWICO debates. Here, public opinion was conceptualized as more than an informed electorate, per Lerner's definition. Effective public opinion implied more socially aware publics in developed countries, particularly in relation to Third World development. However, there was no significant change in the meaning of *development* from the connotations given the term by the dominant discourse. Secondly, information is more closely knit with social communication in this trope, as opposed to a more instrumental, at times depersonalized, understanding of information that was apparent in the dominant discourse. Communication was also conceptualized as a social process, with implications that this move need not necessarily serve the purpose of development. Finally, I examine one document in particular that explicated at considerable length the conceptual underpinnings of alternative media.

Expanding the Concepts of Public Opinion and Information

Modernization debates highlighted the need for full-fledged mass media for building an informed national electorate, and for the purpose of building public opinion in the democratic process. Since Lerner's initial conceptualization of the role of mass media in facilitating public opinion, it has constituted a rationale for development processes. At least two implications were evident in this role for the mass media: (a) that a style of democracy similar to the one in many developed nations was required to make sense of the idea of the informed electorate, and (b) that the media would have an effect on Third World populations similar to that of Western democratic societies. A shift in the attention to public opinion in the First World, and an expansion of the concept from informed electorate to informed publics are apparent in two instances—in the statement of the participants in the Dag Hammarskjöld Third World Journalists' Seminar and in the statement of the goal of NAMEDIA. The Hammarskjöld Seminar participants explicated the need for strong awareness of the idea of development among the people of the economically powerful countries:

> A massive effort must be made in industrialized countries to explain and make known the legitimate nature of the aspirations of the Third World countries for the establishment of the New International Economic Order.... Links with institutions established in the industrialized world that promote, in their own public opinion, the ideas and values of another development for all societies should be created. . . . (Development Dialogue, 1981, p. 117)

In this quotation, the participants expanded the role of public opinion from the idea of an informed electorate to an informed public, and from functioning to support a democratic system only to appreciation of (often) contradictory and conflicting economic and social issues in a globalizing world, of which they are very much a part.

A similar effort to redefine public opinion and its role in a world community is emphasized by Parthsarathi (1983). He integrated the notion of public opinion into the goal of NAMEDIA, which was to create "[a] better informed world community whose members engage in peaceful and creative co-operation for enriching each other through the sharing of human and cultural values" (Parthasarathi, 1983, p. 57). Suggestive as this large statement is of the trope of macrospeak in a different context, nevertheless public opinion of this nature would create a different kind of awareness among publics—one where the "image of the other" would be accepted without "ethnocentric prejudices" (Development Dialogue, 1981, p, 117).

Part of what I read as retheorizing also addressed the meaning, nature, and role of information in communities at all levels. In his presentation at the NAMEDIA conference, one of India's well-known film directors, Mrinal Sen, claimed that communication could only be achieved when the involvement of the receiver was invoked, which is why impartial information, according to him, did not communicate. Instead, he advocated information with comment. This idea questions a tenet of journalistic training that news is objective information intended and designed to communicate facts. Sen's role as a filmmaker, as a communicator, and as a social animal (his self-description) enabled him to envisage and articulate information with comment as a prerequisite for effective communication. Placing ideological conviction before technology, Sen (1983) held, ". . . we need to be partisans. We need to convince ourselves that reality in a sense is strictly partisan. I must interpret it. I must not fail to comment" (p. 35).

Defining information according to a community's needs and identifying "third systems of information" were other means to understand information differently from the issue of news flow balance between the North and the South. The first Director of the Inter-Press Service, Savio (1982), argued that the cause for demanding a NWICO should be based on more than just differentials in the quantity of news flow. He emphasized content, which, according to him, had been insufficiently addressed in the debates:

> . . . is a question of creating new flows of information, with a content, actors, priorities and needs which are absent in the existing flow. In other words the Third World needs to produce information which is not provided by the existing system, but which reflects its realities and communications needs. (p. 75)

Savio advocated the idea of *third systems of information*, which included trade unions, academic institutions, cooperatives, religious movements, and people's associations—systems he understood as constituting the social fabric. According to him, rather than constituting sources for news in the mainstream media, these organizations would be the third systems generating and circulating information. Action decided on in most international meetings ended in the repeated emphasis on technoeconomistic improvements of media infrastructure in spite of the strained national budgets, when what was needed was "alternate projects and research that were also open to third system association." Moreover, gatekeepers in the media of both the North and South needed to be educated about what a new information order really meant, without which distributive justice in information and communication cannot be envisaged[7]

Instead of a NWICO, Savio conceptualized a new information order that was not necessarily international. Constructing a new information order would mean disengaging from the world system, instead of attempting to change it into another one. Second, seeking solutions outside the ones offered by governmental and intergovernmental organizations or the state would remove smaller, local communities from the power influences of these institutional apparatuses and allow them to act on their special communication needs. Third, he also emphasized the need for third systems of information (including nongovernmental organizations) and the importance of funding them. Finally, whether it meant the media of minority political groups, grassroots communities in rural and urban areas, or whatever other definitions described alternate media, Savio (1982) held that "these are (all) means of communication with an alternative content, which by-pass the so-called mass media. This reflects the crisis in the concept of a single medium which can serve everybody" (p. 79). A single medium can be interpreted as the modern mass media in general, where the term mass assumes an undifferentiating connotation and reach. What is evident here is Savio's recommendation for diversifying across space (cultures) and time (context).

Returning Sociality to Communication

In this subsection, I borrow the concept of *sociality* from Appadurai to convey attempts to retheorize communication. Appadurai (1990) defined *sociality* as follows: "Much of what is seen as valuable by members of the community, and much of what appears to underlie the central traditional forms of social life, are linkages between persons and groups, taken for granted not only as means but also as ends" (p. 180). In his anthropological study of

---

[7]For example, Savio observed that gatekeepers continued to value spot news, which in cases of a breaking story acquired tremendous news value for the news organization.

agronomy in rural western India, Appadurai demonstrates the invasion of modernity as an exogenous force altering agriculture and social relations through the establishment of the primacy of the economy. By "traditional forms of social life," we can also read "premodern," a time when economics was integrated into the social practices of village life, rather than isolated from such practices and elevated as a definer of social life. This is apparent in some of the texts analyzed in this section. For example, Parthasarathi preferred the idea of *social communication* to *modern mass communication*. He described the term communication by invoking ideas passed down through the generations (orality), and by exchange of ideas across borders since as far back as the sixth century (e.g., the practice of university exchange programs in ancient systems). Parthasarathi framed communication in the history of international cultural exchange from the early civilizations, heritage, and the social. There is a romanticization of premodern cultures ("cradles of civilization in Asia, Africa, and South America"), but his observation that, with the advent of colonialism and technologies of the North, "our people started looking to the North and the West rather than towards each other" points to the loss of sociality at an international level that was present in an earlier era (Parthasarathi, 1983, p. 55). Further, Parthasarathi held that the "communications revolution" was a revolution of techniques and scale, obscuring in this instance a vital element of the communicative act—content/message.

We return to Appadurai's definition of *sociality*—"linkages between persons and groups, taken for granted not only as means but also as ends"—in reading Tehranian's (1978) UNESCO document on communication theory. Tehranian pointed out that, because of the purely instrumental role assigned to the media in the modernization or development process (he used both terms interchangeably), the

> mediacentric views have ... neglected what seems to be the critical functions of communication in the processes of modernization and social change, i.e., the epistemic differentiation and pluralization of life worlds and the need for more rational and humane systems of knowledge, organization and reconciliation of the conflicting sources of identity, authority and legitimacy. (Tehranian 1978, p. 2)

In the mediacentric views prevalent in the discourse of communication and development, communication becomes a means to attain development. As many have observed, and as Tehranian also pointed out that as a means, mass media growth serves to concentrate power in the hands of a few and allows them to "exert cognitive control and domination over the many." Further, the needs that images of development raise are irrelevant and at times detrimental to the interests of the Third World—"The communication needs of Third World countries for authentic self-development often run counter to the images and imageries of development imposed on them by alien sources." Tehranian (1978) also observed that the displacements effected by an exogenously imposed "historical transition like development . . . has undermined the self-sufficiency of the indigenous economy without providing for productive interdependence; it has homogenized and depersonalized the old cultural patterns without giving a new sense of cultural autonomy and creativity" (1978, p. 3).

This critique about a lack of a new sense of cultural autonomy and creativity has been challenged, as we see in the next chapter. Here, we can pursue Tehranian's concern for an absence of sociality in the dominant discourse of communication and development. To continue, in the international picture, the Third World falls somewhere in the "twilight of tradition and modernity, suffering the worst consequences of both without benefiting from their blessings" (Tehranian, 1978, p. 3).

For Tehranian, the continuity of colonialism in the development project and its effects on communication followed a psychological trajectory: humiliation of traditional societies by the colonizers and the internalization of feelings of inferiority by the colonized led to the persistence of the "slave's acceptance of self-image of depravity and inferiority in the colonial era." The cognitive change of the colonized then gave way to an elite-mass dependency in the postcolonial era (Tehranian, 1978, p. 3). In this context, the First World's treatment of communications as a means to ensure a (an albeit loose) collection of ideas, institutions, practices, and lifestyles representing an end, and the Third World's techno-economic valorization of both the means (communications) and the ends (development) have contributed to diminishing sociality in communications. Further, Tehranian continued, modern communications technology increased the chasm between the idea that communication is a strategic instrument for achieving development and the idea that communication is more an organic social act that cannot be separated from the sociality that is its essential characteristic:

> Technological determinist views of history and a pre-occupation with the big media and its facile channels of access without a commensurate concern with communication feedback and participation will produce only alienation and protest. Communication is essentially a process of humanization and cannot thus be divorced from those delicate and subtle bonds of human society which are called art, language, culture, ethnicity, identity, legitimacy, or ideology. (Tehranian, 1978, p. 5)

Critique of the dehumanizing nature of technology is not new. Scholars from the cultural imperialism perspective have also written about the invasion of communications products of these technologies heavily overlaying and possibly erasing traditional Third World cultures. What differentiates Tehranian's (1978) argument from these others is his non-romanticized and close look at premodern or nonmodern communication as integral to the social culture. He said of modernization: "Modernisation's first task is to re-write history, to obliterate the historical memories which stand in its way. . . . To debunk them altogether in the process of modernisation is to reject the possibility of all communication and all civility" (p. 6). However, Tehranian did not provide alternatives to this situation. There seem to be brief glimpses in his text of a belief in a new modern society more rational than traditional, suggesting overall more humane communication practices than the existing (modern) ones would allow.

Like Tehranian, scholars Martin and Hindley, in their article entitled "A Plea for Humanized Communication," sought to separate the overwhelming influence of technology on communication from the sociocultural nature of communication. Their recommendations may be unrealistic in the present age of broadcasting, but their articulations of what might be termed as the "human" dimension to communication is important. Their primary concerns were with two new technologies—TV and the satellite. Martin and Hindley held that the influences of communications technology resulted in inequality in the speed of communication and, consequently, "destroy[ed] community." They provided some criteria for assessing whether "an advance in communication technology is actually an advance in human communication terms" (Martin & Hindley, 1981, p. 110). First, does the technology facilitate easier access to stored human experience? During the NWICO debates, one of the communications problems identified was the Third World's limited access to computer databases and satellite information that essentially became the property of the industrially advanced nations. A second criterion included the size of the common information space for those using the communication system. According to Martin and Hindley, the standard for such a space was face-to-face communication; they noted that TV failed to achieve this criterion and was not considered to be an advance in communication. More, "it is destroying communication at the level where communication and human interchange is still possible." As the third criterion, communications technology had to facilitate "the discovery of consensus," considered an advance in human communications terms. They pointed out again that TV failed this criterion.

To extend their argument, the TV industry constructs and, on the basis of this construction, assumes viewer consensus in its mapping of target audiences according to numerous characteristics. Martin and Hindley used these criteria to amplify Innis' valorization of oral culture. They quoted Carey: "[Innis] refused to yield to the modern notion that the level of democratic process correlates with the amount of capital invested in communication, capital which can do our knowing for us . . . [replacing the] political power of foot and tongue" (Martin & Hindley, 1981, p. 108). Thus, dialogue, rather than one-way satellite communications, would facilitate a culture's understanding of its own position in the contemporary world.

In an analysis of the adoption and adaptation of modern communications technology within a non-Western culture, Komatsuzaki demonstrated with the case study of Japan that for sociocultural acceptance of a new communications technology, it was crucial to adapt that technology to the host culture. In a UNESCO report entitled "Communication Technologies of the 1980s: The Social Implications," Komatsuzaki (1978) identified links between cultural specificity and parameters and acceptance of a new medium. Using the case of Japan, Komatsuzaki explained the complexity of written Japanese and demonstrated how children's communication behavior changed with the advent of the telephone. Among the young, telephones became the preferred mode of communication displacing forms of written communication such as letters. He also pointed out the adaptation and streamlining of the Japanese script to the computer keyboard. Komatsuzaki's data on the industrial, urban, youth, and other demographic profiles matched similar newer media use trends among the older, retired populations who sought leisure from private spaces like the home. Home computers would also facilitate what he called the *electronic learning society*.

Komatsuzaki's analysis pertains to Japan, an economically powerful and technologically advanced nation, but some openings for consideration for the other are evident in his document. Komatsuzaki (1978) warned that, prior to adoption, "social acceptance of new experimental media depends on their ability to . . . conform with the regulations of the culture involved" (p.1). He emphasized the importance of the context in which modern communications technology would be received by host cultures; the implications are that if the goodness of fit does not occur, that technology is not suitable for that culture.

Attempts to return sociality to communication assumed many guises. From Parthasarathi's efforts to connect communications with history to Tehranian's, and Martin and Hindley's articulations of a more humanized communications, to Komatsuzaki's emphasis on the receptive conditions of a host culture for new communications technologies, attempts were made to define *another development* through *another communication*, using broader sociocultural and humanistic rather than specific economic and technological bases and criteria for establishing development through communication. A progressive agenda and a strong assertion of the Third World presence are evident in attempts to return sociality to communication. Also evident are more radical political projects of alternate communication, such as the one in the document discussed in the ensuing section.

Alternative Communications

In the last few subsections on the trope of retheorizing, a reading of the texts demonstrated the conceptual underpinnings for another understanding of the practice of communication. Such conceptualizations drew from existing ideas of modernity (Tehranian's new modernity), flexibilities offered by satellite technology (Pal's strategies and models), cultural fit for housing technologies in developing societies, and connections of communications to history. Although these conceptualizations sought to interrogate the treatment of communication in the discourse of communication and development, they were less focused on articulating alternative communications that could simultaneously contribute to and emerge from social transformations. Alternative media and alternative communication is now extensively historicized, theorized, and practiced. Seminal works in recent years, notable among them Atton (2003), Downing (2001), and Rodriguez (2001a), to name a few, have addressed the range, motivations, and politics of alternate media. In this section, I discuss one UN document as an example of a counterdiscourse to mainstream meanings of development and media. This text is an example of alternate communication written during the NWICO debates dealing explicitly and extensively with alternative communications.[8]

In his article entitled "Alternative Communications," Matta (1979) defined *alternative communication* as "a social process whose contents and meanings are determined by the political activities and experiences of dominated sectors" (p. 98). These sectors could comprise anywhere from marginalized urban groups in individual First and/or Third World cities to marginalized groups of nations. Matta understood alternate communication to be a cultural reaction to economic, political, and cultural disenfranchisement of such marginalized peoples. Here, communication constitutes a process that is an experiential outcome of such groups. He identified an overall role for alternative communication as "protect[ing] and develop[ing] national cultures dominated by transnationality which hinders and destroys their independence" (Matta, 1979, p. 96). Both the economic and cultural dependency critiques are apparent in Matta's statement. This article implies the existence of an authentic national culture and the importance of securing this culture in the realm of alternative communications.[9] The danger was in transnational communications' imposition of an alien framework of reference for social activities, political values, and public opinion. Alternative communication would resist such (total) impositions. Alternative communi-

---

[8]An extensive treatment of participatory communication as an alternate form of communication can also be found in O'Sullivan-Ryan and Kaplún (1980) during the period of the NWICO debates.

[9]The authenticity of the concept of a national culture and identity has been heavily debated. The literature includes scholars as diverse as a major proponent of new communication technologies, Pool (1990), and Marxist historian, Chatterjee (1993). I pursue the point of essentialist conceptions of identity to the extent that it enables my discussion of the popular and hybrid in the next chapter; more extensive discussion is beyond the scope of this chapter.

cation is conceptualized as having a dialectic relationship with transnational communication, particularly because the dominant model of communication "offered by the world oligopolies is unidirectional, vertical, imperative and not chosen by the people," and alternative communications is a more horizontal (accessible to the mass), popular practice (Matta, 1979, p. 97).

Matta dichotomized the world order into two poles—the dominant and the dominated. Although his discussion suggested that the dominant pole represented transnational communication, and the dominated pole (the Third World) constituted more local popular or vernacular communication, he included dominated spaces in the First World also. This inclusion is reflected in his conceptualization of a horizontal field of communication. It is the power of the other that marginalizes people and consequently engenders alternative communication. Thus, "the alternative communicator provides an outlet for the symbols, meanings, aspirations and struggles which motivate the people of the dominated pole" (Matta, 1979, p. 99).

According to Matta, a democratic field would be a fundamental prerequisite for alternative communications to challenge the power of transnational communication, and it is in this field that "the emerging ideas and historical projects of a given society are confronted" (Matta, 1979, p. 98).[10] He extended this democratic field to encompass the international arena—his conceptualization of the "true dimension of horizontal communication." Alternative communication experiences could be shared, through transfer, "between the Dominated Poles of different countries [First and Third World] which coincide in their struggle to create alternative communications in opposition to their respective Dominating Poles" (Matta, 1979, p. 101). This sharing would engender a "power which the creation of authentic social awareness generates," presumably to effect social changes suited to the context and realities of various communities and societies.

Matta demonstrated the potential for effective alternative communications with four arguments, and key reasons that included fluency in the vernacular/spoken language, shared social experiences, and participation. First, he emphasized the high source credibility in the communicative act—"the contents of alternative language are generated by a specific shared social experience that the audience can readily identify and respond to" (Matta 1979, p. 99). Second, language emerging from a close sharing of experiences would permit a decoding that would be both fluent and "highly coinciden-

---

[10]Martín-Barbero (1993) took the idea of the resurgence of interest in the popular a step further, extending Matta's engagement with historical projects. He saw this resurgence of interest as bringing the popular into mainstream cultural history. Matta's article was written in 1979; since then, Third World societies have taken what has been termed elsewhere as a postdevelopment turn (Escobar, 1995; Rahnema, 1998). This turn is reflected in Martín-Barbero's work cited here, which was first published in 1987; the English translation was available in 1993.

tal with the purposes of the emitter," and would also create "optimal conditions for feedback and participation." Third, the shared alternative language could intensify communication experiences. Fourth, the motivating potential of alternative communication is high for whatever change the community may decide on. Some alternative communications media include, but are not limited to, graffiti ("writing on city walls"), traveling theater, news bulletin boards, and what Matta termed "other marginal forms of expression." Alternative language could also use communication avenues more characteristic of the Dominant Pole—films, video, radio, forms of recorded music compilations, and so on. But the fundamental difference

> . . . lies in the "recodification of their use," which differs from the way in which the industrial system intends them to be used. Consequently, the difference between mass media serving cultural industry and alternative media lies in the access (or lack of access) to certain media by groups which generate alternative language, and…introduce alternative contents. (Matta, 1979, p. 99)

Thus, access, one of the primary concerns in the NWICO debates, is addressed here through participation, shared experiences of struggles against transnational power, and the bonds of the vernacular. From access, Matta then shifted his arguments to flow—communication among the groups of the Dominated Pole, against and beyond the Dominant Pole, would ensure a more equitable distribution of power. His case study of the Chilean musical form "Canto Nuevo" and its circulation testified to the power of alternative communications and its ability to alter the direction of communication flow. The Canto Nuevo form emerged after the military coup of 1973 and encoded sufferings of oppression under the Junta regime in its lyrics. These songs were sung in locales such as solidarian meetings, festivals, and in other popular locations frequented by the vast numbers of marginalized peoples, where alternative communications would be most effective. Eventually, however, this mode of horizontal communication was incorporated into the national broadcasting system (the ownership of which included foreign interests), and was met with much enthusiasm by recording companies. Although the trajectory of this popular musical form from the grassroots to transnational circulation technically altered the directionality of the flow, Reeves' analysis of the transnational music industry in the 1980s and the early 1990s is a reminder of the negative effects of cultural flows in this direction. With examples from West African and Caribbean popular music such as reggae, Reeves (1993) noted that ethnic and indigenous forms of music were incorporated either wholly in the transnational music companies or were used in parts to revitalize existing markets in the West.

Another understanding of the concept of development emerged from efforts to re-theorize communication. What was hitherto seen as an ultimately economic enterprise was now being considered as cultural emancipation. Traces of a shift were also evident from the social engineering of societies and the masses to a self-constitution at the popular level through popular/humanized communication. In Matta's article, in particular, what emerged was the idea expressed later by Martín-Barbero (1993) as "the people as the subject and protagonist of history, until recently denied by a historical method that buried the people under the epitaph of an autonomous mass of numbers" (p. 62). In the next chapter, I develop this idea further to examine later stages of communication and development.

CONCLUSION

In this chapter and the last two, we have seen how even among the proponents of a new information order, divides were apparent. Resistance to and negotiations with the discourse manifested themselves in several degrees of intensity. From the proximity to the center, alternate uses for technology were suggested. The other end of the continuum favored a break from the center to establish popular voices in communications and popular culture as a manifestation of the quest for social change. Limits to the expression of resistance in the texts discussed in this chapter, particularly those proposing a change that conforms more to the mainstream notion of development (although refusing to imitate the First World) are apparent. In the case of restructuration, we can see the limits as Layder (1994) expressed them:

> . . . people and the products of their social activities cannot be treated [independently]. People are intrinsically involved with society and actively enter into its constitution; they construct, support and change it because it is the nature of human beings to be affected by, and to affect, their social environment. (pp. 127–128)

To the extent that the world order emerged as a historical development, supporters of a new information order could express the felt need for different ways of conceptualizing the world order, communication, directionality of communication flow, and so on within the legible metaphors of the existing world order. As Huesca and Dervin (1994) have explained, alternate communication can be recognized as such within specific contexts, taking into consideration its inclusion of the popular.

In the next chapter, I discuss a trend toward viewing and practicing communication as an expression of culture, identity, and as particular to a community that was present while the NWICO debates were also under way. Much has also been written about new directions for communications and development, among them works addressing social movements and their shift from an institutional to a popular cultural realm. I draw partly on this literature to demonstrate the idea of communication as human praxis (action for social change) as it occurs in the context of local knowledges and practices.

5

FROM COMMUNICATION AND DEVELOPMENT TO POSTDEVELOPMENT AND GLOBALIZATION

Debate on the meaning and practice of democracy continues to animate discourses on communication. Theories and practices associated with the various stripes of democracy are manifold, but certain fundamental characteristics appear consistently in the larger political conversation. For example, participatory democracy prescribes a voice and access to voice for all citizens in a polity. Representative democracy functions at the governing centers of a state through elected representatives. There is also the democracy associated with capitalism, where markets are unfettered by state rules and are free to respond to the forces of supply and demand. Like democracy, development has an equally complex genealogy. It is at once a group of theories, a lived reality, a materiality, and a discourse (Apter, 1987). "Developmentalized democracy" (Escobar, 1995) is the product of capitalist democracy, a dominant discourse of development, and a brand of democracy that is observable in the NWICO debates.

In this chapter, I chart the move from the concept of the international to the concept of the global, locating communication, development, and social change within this shift. This movement from the "international" to the "global" also emerged from the NWICO debates. A link to what has been described as the postdevelopment era lies in the current understanding of the phenomenon of globalization; I discuss the move from communication and development to culture and globalization in the ensuing section. The postdevelopment era is one of hybridization (in practice, and also in the search for appropriate theories), in which context grassroots level and popular communication efforts in developing regions have been increasingly documented. These communicative practices are directed more toward kinds of social change resulting from the communities' own choosing and response to their particular histories and contexts. In an upcoming section, I present some examples in this area to demonstrate these developments. Finally, I provide some observations about what the documented research on grassroots cases might suggest for global or horizontal connections that are hinted at in some of the texts read for the tropes of negotiation and resistance.

FROM COMMUNICATION AND DEVELOPMENT TO CULTURE AND GLOBALIZATION

Since the withdrawal of the United States and Britain from the UNESCO, mixed opinions are apparent in the evaluations of the NWICO debate—that the issue of a balanced information flow is no longer of consequence, or that it is apparent in other discourses such as intellectual property rights (Braman, 1990). Studies such as the one by Boyd-Barrett and Thussu (1993) evaluated the performance of Third World regional News Exchange Mechanisms around 1993 (more than a decade after the publication of the MacBride Report) and pointed to the enduring nature of the issues raised in the NWICO policy debates. The discourse of communication and development continues to have much appeal and is evident in aid projects that are sponsored by agencies such as the US Agency for International Development and UNESCO.

A collective phenomenology has now arrived at globalization as the latest development in the international space (Tomlinson 1996). Concerns about culture, identity, and the global/local are one set of issues that are now being avidly debated by sociologists, anthropologists, developmentalists, political scientists, and economists. Yet, as Sreberny-Mohammadi (1996) pointed out, these concerns have been present in communication studies for at least a quarter of a century now. Hamelink and Pavlic, Schiller, and others have dealt with transborder information flow and the control of the flow by Transnational Corporations (TNCs); the crux of the problem lay in the TNCs' bypassing of state control in their operations. Nevertheless, Robertson (1992) held that an intensification of globalization, both in cultural and economic senses (Appadurai, 1993; Lash & Urry, 1994, and others now refer to a global cultural economy), is acutely felt in the early 21st century.

Current preoccupation with the phenomenon of globalization among many academics and policymakers alike suggests that we are moving into a new world order that is gradually rendering our understanding of the older international order defined by the signifier development irrelevant to the present changing conditions. New information and communication technologies are the most frequently cited cause for the occurrence of this phenomenon of high globalization at the turn of the millennium. For example, Lash and Urry (1994) noted that fiberoptic cable, satellite communication, and air transport have affected globalization (as differentiated from wire cable, postal services, and road networks associated with nationalization). The latest wave of new communication technologies has facilitated the spread of capitalism and the elision of national borders by accelerating industrial and financial capital transactions on an unprecedented global scale. The global whole has acquired a new magnitude and ubiquity. Given this recent turn, it is important to understand the continuities and changes between the international and global eras from the standpoint of communication and development. It is helpful to sort out these continuities and changes because they constitute the historical context within which to conduct research on the role of communication for social change in a globalizing world.

We now have to contend simultaneously with the enduring international hierarchies—a legacy of the international development era and, if traced further, the colonial era—and rapid changes that challenge our understanding of this international space. Hence, our present context generates some issues critical to understanding the nature of the shift in concerns from communication and development to globalization and postdevelopment. A fundamental issue points to the relevance of development as a political, economic, and cultural process in a changing time when culture and consumption predominate discussions of globalization. Second, the effects of globalization on the communications needs and actions of the developing world need careful examination. To explore these issues in depth, it would be helpful to investigate limits to claims of difference between the development-based international order and the present global order from the standpoint of communications. It is important to begin to address these issues because the geographical sweep of the current manifestation of globalization could obscure the real hierarchical differences among various regions worldwide. The paradoxical nature of both the appealing simplicity and the inherent complexity of globalization can cognitively distance development questions from the pleasures and advantages of global communications. When problems introduced by the development project persist, and are even exacerbated now, factoring in the development question in the global era becomes critical for knowledge and praxis oriented to social change.

Theorists of globalization emphasize western capitalism as the driving force behind this phenomenon (e.g., Schiller, 1989; Thomas, 1996). The fall of Soviet socialism has been decried as a failure of noncapitalist systems and is seen as signifying the (expected) triumph of capitalism. The growth of certain types of industries, new locales of production, and increased and new forms of consumption mark the process of globalization. Free trade and the market are considered to open up opportunities to all strata of societies. At the same time, culturally, consumers of images and designs (rather than the products themselves—Lash & Urry, 1994) are considered to be far less vulnerable to the homogenizing influences of globalization than critics of the phenomenon would have us believe (Tomlinson, 1996). Skeptics and critics of globalization, however, have different and differing views, some strands of which bear sorting out here.

Tomlinson, (1996), after Giddens, analyzed the global experience as a "consequence of modernity." For him, modernity is ". . . an ambivalent experience of exhilaration, the realization of potential, and a certain precarious control combined with risk, insecurity, powerlessness, and existential anxiety" (p. 63). Critics of modernity/modernization (often used interchangeably in the context of international communication) have held that this is not universally true. A pioneer in the theory of globalization, Robertson (1990) emphasized that, while gathering sweep and momentum and acquiring inevitability in history, the phenomenon is not necessarily a good thing. Critics like Thomas (1996) introduced globalization as a "renewed global organization of inequality" (p.1). Their critique of the phenomenon of globalization can be understood in terms of two types of dislocation: in the center-periphery arrangement, and in the export of problem images from the centers to the peripheries. The first type of dislocation is apparent in the analysis of the TNC takeover, where both people and their governments (states) are cast as the marginalized parties (Bello, 1996; Thomas, 1996). In this frame, both popular and state challenges to globalization are apparent. For example, the then Newly Industrialized Countries (NICs) of East Asia are cited as instances of state-assisted capitalist economies, resistant to the idea of a fully market-regulated capitalist economy. To add a shade of complexity to this picture, there is the idea of sustainable development, where there is opposition to both free trade and state-assisted models of capitalism in the NICs. This issue of sustainable development was first raised by grassroots movements and non-Western Nongovernmental Organization (NGO) critics (Bello, 1996). Eventually, sustainable development was absorbed into the dominant discourse of development (now visible in the ecological issues and related regional trade agreements).

Development continues to be a deferred ideal for many societies. The perceived attractiveness of the still-elusive end state of development has added to its mystique, and for many developing regions is still a desirable goal to pursue. Initially, the final destination of a development trajectory was envisaged as a social order similar to that of modern advanced societies. It suggested a homogeneous formation of an exemplary society. But vernacular modernities have emerged, and internal differences and resistant discourses continue to inhabit the development discourse. When numerous development projects were implemented over the years, variations of modernization changed the idea of a single destination for less developed countries. However, basic ideas about technologies, the economy, and related social institutions underlie these variations. For some scholars, development is part of "the larger history of Western modernity" (Escobar, 1992, p. 22; see also Esteva, 1992).

Clearly, some changes to this picture are evident from the last decade or so. The main causes attributed to this change are two interrelated factors: (a) the failure of development projects and ensuing frustrations at the local level within nations, and (b) globalization. The global turn toward the final decades of the 20th century accentuated informational, economic, and technological disparities on the one hand and on the other, created new opportunities for mobilizing for social change. Chasms caused by technological disparities on the whole exacerbated ecological and gender disparities, posing new development problems (Shiva, 2000). Martinussen (1997) pointed out that globalization has involved "a much more profound reorganisation of manufacturing, trade and services" (p. 120) when compared with the international era, although the latter did involve an extensive spread of production and manufacturing to many parts of the world. New developments in communications technologies and new spaces for communication like the Internet have served to equate the term *global* with its literal interpretation, worldwide. Communication and information have played a pivotal role in internationalization as well as globalization.

The developing regions play a dialectic role in the phenomenon of globalization, both in accepting free-market imperatives proposed by institutions controlled by Western powers as well as by opposing global and state-imposed development models on cultural grounds—in many instances, at the grassroots. That the state facilitates free-market imperatives forms another critique of globalization (Shiva, 1997). This critique leads to the ways in which communities at the grassroots cope with the phenomenon of globalization. Saukko (1996) observed that problems which arise out of globalization are "mostly fought where they are mostly felt . . . in the local," Uneven globalization results in differing responses to the phenomenon (Thomas, 1996). Hence, Although globalization can be understood as a threat to the local from an argument driven by capitalism and a cultural economy, the effects or threats felt by local spaces seem to vary. An alternative picture of globalization emerges from building communities across borders. The ideology of culture and globalization seems at various points to replace or coexist with the ideology of communication and development. Expectedly, resistance to and negotiations with the process of globalization, as for development also, was varied both at the state and community levels (Sosale, 2002).

Increasingly, research has documented the developing world's protests against the dominant notion of development and the social and material conditions that have for so long defined it. The volume of documentation has increased now, but these expressions of dissatisfaction in the form of social movements have existed for some time. A loss of history and identity has been felt in the process of undergoing mainstream development changes, and development has come to be perceived as a hindrance rather than a help to social change. Legitimation of a certain type of economy, state, and society has, in Mohanty's (1991) words, "suppress[ed] . . . material and historical heterogeneities" (p. 53), which struggle to find expression in the present-day developing societies. Such expressions have been articulated by and have mobilized grassroots movements of rural communities and the poor of the urban centers. In this sense, the disadvantaged populations in the First World have been defined as part of the Third World as well. In this context, multiple communicative experiences signify Third World—expressive practices of the rural Third World, of the urban poor in the periphery, and of the economically disadvantaged in the center (Chanter, 1991; Williamson, 1991). The latter two are referred to as peripheries, in the plural.

Escobar (1995) observed that development has meant "choices and power of some and the chances of others" (p. 215), and that the majority of the world population is part of a world economy that has little of this population's interests in its operation. Consequently, the majority is "marginalized from its benefits but is integrated into its effects" (Escobar, 1995, p. 222). Increasing instances of documentation demonstrate that local knowledges and local power have acquired significance, and the quest now is "not [for] development alternatives but alternatives to development" (Escobar 1995, p. 215). About a decade earlier, Rist (1997) made a similar observation: Either alternative strategies would lead to other kinds of development or new definitions of development would demand alternate strategies. In either case, as Rist pointed out, the "necessity of plurality of developments is becoming more and more recognized" (p. 102; see also Development Dialogue, 1979).

According to Colombian scholar Martín-Barbero (1993), state impositions of a national culture on heterogeneous citizens in many countries of Latin America are similar to the transnational impositions of limited ideas of development in the Third World. The situation is complicated by the peripheral state's connections with the center and the influence these connections have on the idea of a national culture, and also by colonial history and forced migrations that have rendered the population very heterogeneous, and thus consistently problematic for the idea of a national culture. Hence the increasing power and appeal acquired by the popular and localized identities, a result of conflicts "which are situated at the intersection of political culture and the new understanding of a cultural politics" (p. 453). Melucci (1990) observed that the act of naming was akin to "bringing into existence"; literature has demonstrated that colonialism and later modernization deprived large numbers of people of this "power to name."

Consequently, mainstream modern mass media's labor to interpellate a public gives way to a public that is fully involved with the communicative act. Under these conditions, democracy would not be "just a question of altering structures by a legislative fiat but [would] itself [be] a method, a process of communication" (Kothari & Sen, 1983, p. 11).

In his reconceptualization of communication and development in the context of the NWICO debates, Tehranian (1978) characterized developing societies as caught between the ties to the traditional (and the known) and the angst of modern societies.[1] This angst may point to the fears of the state and the urban middle class in many of these countries, but we know that communication among the disenfranchised may show otherwise. Tehranian's reading of both the traditional and modern defines boundaries for meanings of communication and development. Metropolises in developing societies now have multiple pockets of solidarity and identity, and the anomie produced by modernization and its requisite displacements has reorganized itself into the popular. Evidence also exists of rural societies evolving coping, resistant, and other strategies for survival, adaptation, and self redefinition in the face of globalization. Huesca and Martín-Barbero, among others, have questioned the antinomy of tradition/modernity from a critical perspective, and have demonstrated instances of hybrid communicative practices from the standpoint of the peripheries. They also demonstrated the constant negotiations between centers and peripheries, thus recognizing the changing and dynamic nature of communication emerging from the popular. As Escobar (1995) explained, "traditional cultures' transformative engagement with modernity incorporates transnational motifs in traditional designs" (p. 219). A specific instance of this incorporation is presented in the example of a study of folk theater in rural Kenya in the following section.

Perceiving that the practice of popular participatory culture has overtaken efforts to build conceptual frameworks for understanding it, Huesca and Dervin (1994) critically documented the practice of popular culture through the vital dynamic of participatory communication.[2] For Huesca and Dervin,

---

[1]Tehranian (1994) has since expanded his conceptualization of communication and social change. This maturation is reflected in his chapter entitled "Communication and Development."

[2]Critical observations of the idea of participation in development projects, such as Rahnema's (1992), hold that participation is an inherent activity of traditional, rural societies, and therefore there is little use in introducing the idea of participation in these social contexts. I agree with Rahnema's critique of the discourse of development that participation in developing regions is insufficiently recognized, but I would qualify it with "in many instances." Women's conditions in some societies where opportunities for participation have been traditionally nonexistent come to mind. For example, research indicates that women in rural North Indian communities are one of the most difficult target groups to reach for participatory communi-

the term *popular* means "that which makes possible the expression of collectively produced expectations and aspirations of and by popular social groups" (1994, p. 60). The expressions of these social groups occur through alternative communication. The popular is a space where "social contradictions, arising from national or other transnational cultures that are imposed from the top and the local cultures that resist from below, are symbolically resolved." (Huesca & Dervin 1994; they also cite García Canclini, 1988). Symbolic resolution occurred with and through a synthesis of the traditional (such as orality) and the modern (modern mass media). This synthesis was reflective of a combination of political awareness of marginalized positions, the realization of the importance of everyday life (which partially explains that resistance at the level of the popular need not necessarily be intentional), and an intense need for expression especially under repressive regimes, but also as a response of the peripheries to the centers (class).

Huesca and Dervin's (1994) concern was with alternative communications leading to social change; they looked to what they called "alternate media theory" for treating communication as a "tool of a larger social praxis" (p. 57). They differentiated between alternate uses of the mainstream media and alternate media. This differentiation seems to be problematic—it is conceivable and has also been proved that both types of media could serve in different capacities for groups in the peripheries. According to Huesca and Dervin, mainstream media fragment audiences, whereas alternative communication practices are more concerned with building connections. In the case of the former, people become isolated viewers and consumers, in the case of the latter, they become listeners and viewers. Active audience theories might contest this position on grounds that audiences of mainstream media are also listeners and viewers with intricate interconnections among themselves that have yet to claim sufficient research attention. The differentiating factor in Huesca and Dervin's argument is the conscious effort exerted by practitioners/producers of alternate communications to create connections among the audience members and between themselves and their audiences for the purposes of praxis. This observation is substantiated with the example of the study of the Bolivian tin miners' radio (Huesca, 1994) in the next section. Some cases where alternative communication has been

---

cation related to reproductive health issues because of their infinitely subordinate position in the community (information gathered from conversations and research readings at the Center for Communication Programs, The Johns Hopkins University, 1994). In such extreme cases, some exogenous intervention to instigate participation and empower the women with tools they recognize they can generate from their own conditions would be of considerable help. Huesca's documentation is of a participatory communicative practice that has emerged from within the community, and not an exogenously introduced model by the state or other members of what has been called the development machinery.

observed in action and has contributed to social change will help ground the idea of the multiple "others" that the dominant discourse of communication and development seeks to overcome.

POSTDEVELOPMENT? COMMUNITIES, SOCIAL MOVEMENTS, AND LOCAL MEDIA

Several excellent studies of alternate media and communication in the context of developing societies illustrate the almost organic nature and use of the media by grassroots social movements (e.g., Huesca, 1994; Rodriguez, 2001a, 2001b). I present two examples in brief to demonstrate the politics of alternate communication in specific contexts. The first involves a community in Kenya (it occurred and was documented during the NWICO debates) and the second a rural women's movement in India (which took place in the early 1990s when the process of contemporary globalization was already well on its way). The examples illustrate uses of popular theater, print, and film media. The media content varies and is heavily influenced by the local, in keeping with the culture of the producers and consumers of these messages. The underlying commonalities in these examples are participation, political consciousness, and local cultures finding expressive outlets in the media (i.e., instances of democratic communication that are not necessarily tied to institutional definitions of democracy through development). The first instance illustrates the use of popular theater for a social movement in a rural community in Kenya and the subsequent move of the performances to the capital city. The ways in which the margin defines itself against and away from the center become apparent in the peasants' own (alternate) definitions of *literacy, participation*, and *mass*. Next, a documentary video on a women's social movement in India provides an opportunity to study the role the media play in providing impetus for the movement, as well as in disseminating the efforts of a rural population to global audiences.

I draw from Fiske's (1993) conceptualization of local knowledges, which provides a lead for understanding the hybrid nature of media activities and content in alternative media. I also draw from Slater's (1992) analysis of development from a postmodern perspective to explain the specific groups responsible for generating alternative communications in these case studies. Local knowledges are defined by their relationship to knowledge produced in the center, as they in turn define and influence knowledge production in the center. For this reason, local knowledges are multi-accentual (Volosinov, 1984), often incorporating aspects of both the traditional and modern. Slater observed that, to engage in the politics of social change, emerging groups fragment the limited social collectivities of class and the rural/urban. The outcome is that these "struggles for democratic transition take many forms, connecting inter-alia with ecological, women's, urban/ethno-regionalist and human rights issues" (Slater 1992, p. 311).

Example 1

Documented by Kidd (1983), the Kamiriithu Community Educational and Cultural Center (KCECC) in Kamirrithu, Kenya, emerged from a complex history of colonialism and cultural shill economic neglect by the then postcolonial Kenyan government and its affinity with transnational capital, a welling up of villagers' frustration at poverty, unemployment, and suppression of traditional village culture, and the initial organizing impetus for the establishment of the KCECC provided by intellectuals from the University of Nairobi. The impetus for action and cultural production derived from the local literacy program that "encouraged people to question what was happening to them," in contrast to traditional literacy/development programs that accepted the existing living conditions of the people (Kidd, 1983). The KCECC was first established as a people's community center for adult education classes and recreation. Decisions concerning the KCECC were required to be approved by a board elected by the members of the community. The *littérateur* and professor from the University, Ngugi wa Thiong'O, and a literacy program worker from the same university, Ngugi wa Mirii, both driving forces, no doubt influenced the structure of KCECC.

Community members met at the KCECC to decide their study program; this initiative rested on the premise that although formal literacy was lacking among the majority of the population in Kamiriithu, their life experiences gave them a different literacy that served as a resource for teaching and learning. Basic problems of the village such as land tenure (an aftermath of colonialism and colonial agents' involvement in land reform), lack of access to water, high prices of food, and other issues constituted the curriculum. The literacy program proved successful, and the community decided to stage a play drawing on the newly developed resource of literacy. Kidd (1983) described in detail the highly participatory nature of writing the script, raising funds to build an open air theater, selecting the cast, and staging the play. The play, written by "the two Ngugi's," combined material from discussions at the community center and the autobiographies written by students of the literacy classes. In this way, it was possible to incorporate villagers' histories, struggles, experiences, and hopes. Religion, class, gender, and colonial collaborators were addressed in the play.

The success of the play attracted nationwide audiences. Eventually the government, through the District Commissioner, withdrew the license for the play on the charge that it "fomented strife between classes." Ngugi wa Thiong'O was imprisoned, to be released a year later by a new regime. The national press played a significant role in publicizing both the success of the play as well as the forced closure that followed. Nationwide public support was evident through a flood of protest letters. Although Kamiriithu demonstrated strength and resilience in the face of opposition, the license withdrawal affected what might potentially have been an inspiration to other villages in the country to initiate similar literacy and, possibly, resultant popular cultural efforts. A new literacy program was initiated, and a second play was written by the KCECC, with the active participation of women's groups this second time. Attempts to take the play to Nairobi were thwarted by the government, the license for KCECC was again withdrawn, and the village theater was dismantled.

Besides demonstrating the participatory nature of the KCECC's organization, financing, and performance, this case highlights certain other issues. First, the traditional and modern were combined to produce a result that was unique to the context, history, and culture of the people. Traditional folk songs were incorporated in the play as were the life and colonial histories of the people. The influence of the modern is also evident—a faculty member from the University of Nairobi initiated and wrote the script for the play. Second, links to historical sources of present problems appealed to audience members outside the village and evoked a common response to both colonialism and neocolonialism. Finally, the outcome of this experience was a record of popular culture highly political in nature, with the intention of conscientizing the attending public. The efforts of the KCECC were aborted, but persistence beyond the first shutdown and Kidd's (1983) documentation indicated that "awareness, commitment and organization will produce new struggles and new forms of protest" (p. 25).

Example 2

In the early 1990s in the state of Andhra Pradesh in India, the triumph of a women's movement in villages statewide marked a legislative change at the state government level. A documentary was produced to communicate this movement to audiences in India and abroad.[3] The women's movement was aimed at eradicating the problem of state-licensed distribution of indigenous liquor in the villages and the attendant poverty and domestic violence in these communities. The documentary explored the context within which the

---

[3]The documentary video, *When Women Unite: The Story of an Uprising*, was screened outside India for the first time at the 25th annual South Asia Conference, University of Wisconsin, Madison, Wisconsin, USA, 1996.

movement occurred by re-creating the event through reenactment, narrative, and dramatization with the help of movement participants who agreed to act in the film.

As with the case of the KCECC, the crux of the movement lay in its literacy program. Although the source of the curriculum for the program in these villages is not clear, the curriculum served as a conduit for exploring the devastation that alcoholism could wreak in a community. State-licensed distribution of indigenous liquor, controlled and consumed almost entirely by the male members of the villages of that state, was pointed to as the root cause of several social problems. The problem acquired proportions of such magnitude that all forms of income, including women's wages eked out from work in the fields (by this time the only income for the family), would be appropriated by male relatives for purchasing alcohol. The dramatization in the documentary substantiated the effects the women suffered while working to protect livelihood, children and their education, their own work in the fields, and their households. They included extreme poverty, and domestic abuse and other forms of violence perpetrated against the women. With the help of literacy classes and strategic members of NGOs, women in the village of Nellore worked to eradicate the problem through various methods. Efforts to organize and act were the women's initiative; intervention by external agencies such as the NGOs helped the village women deal with local and state governmental bureaucracy. The state and local press covered the movement at various stages; the newspaper reports would be read and cheered by groups of literacy class students (members of the social movement). The women followed closely the spread of the movement as it was reported in the local and the state press. With the spread of the movement statewide, the women's demands on the state government to ban liquor distribution through the state government outlets grew increasingly vocal. At election time, the women used the chief minister's bid for reelection as a strategic lever. Threats to withhold votes if the ban was not issued were followed by a large female electorate that refused to attend elections until their demand was met. Because this electorate would have significant impact on his reelection, the chief minister relented on the eve of the election day and the ban was issued.

The problem of the women of Andhra Pradesh may not appear as a direct development issue. The proximity of the issue to their survival prompted action for social change, such as protest, night vigils, and intimidation of those who delivered the liquor to the village, all of it with little financial aid. Aid appeared in the form of literacy programs, occasional intervention by NGOs, and reports in the press that spurred the movement further through its role as motivator; the movement gained momentum through increased numbers of participants. In showing the emancipation that the victims effected for themselves and their winning the battle with the government in and through the election process, the filmmakers documented an instance of democracy and social change. The makers of the documentary played a large role in disseminating information about the movement among wider audiences. The two women filmmakers included a graduate of a Development Communication program and an economist, both trained at U.S. universities. The film about the movement and the press in the movement served, at different levels and at different times, to disseminate the voice and agency exercised by a population often treated as particularly disenfranchised — rural Indian women. The company behind the production, the Drishti Media Group, is dedicated to publicizing social movements for broader public support and, in the process, making communications efforts "people-centered rather than technology-driven." They explained their methodology as follows:

> People are not reduced to passive objects of our work, whose knowledge, ideas and emotions are utilized without reciprocity. We believe that they can be active subjects in the process of documentation of expressing their own life experiences. This has often taken the form of them writing scripts and acting in the films/plays we make with them. The video films/plays that emerge from such a collaboration must feed back, enhance and support their social and development efforts in creative ways. (The Drishti Media Group 1995)

The combination of technology and participation, in this instance, points to development communication of a kind where agency in struggle for social change is presented as a democratic process unfolding over the course of a movement.

COMMUNICATION AS HUMAN PRAXIS IN GLOBAL CONNECTIONS

Pieterse (1990) defined *emancipation* as "the self-liberation of the non-privileged" (p. 8). It suggests a confluence of the "negative moment" and the "creative moment," where a transformative politics is born. From the potential offered by the tropes of resistance to the examples in this chapter (by no means exhaustive, only illustrative), we see that increasingly there has been a move beyond the dominant meanings that communication and development have been conferred in the international arena. Subaltern groups have created new spaces with the folk, traditional, and modern modes of communication, and these groups have been able to express themselves to small and large publics. Martín-Barbero's (1988) observation that mediation is replacing the media is illustrative of the social nature of communication apparent in the earlier examples because it links members of a society in almost all aspects of the production, mediation, and processing of messages.

The question would then arise as to whether pockets of localized communication would be the solution to problems in the dominant discourse of communication and development. Practitioners, activists and scholars defend such fragmentation as legitimate provided there is political awareness and, equally, tolerance of the other(s) among all social groups. The rhetoric places itself in a realm beyond development, and it is evident in the works of development critics such as Esteva (1996), whose essay was entitled "Hosting the Otherness of the Other" to illustrate both reflexive awareness and acceptance of difference. Under retheorization, we find some openings to establish global links of a different nature—one that is removed from the discourse of communication and development. Whether the materialization of those suggestions would guarantee the tolerance that postdevelopment scholars ask remains to be seen. But the idea of linking to share experiences through communication also demonstrates the possibility of a loose coherence of societies that might contribute to a more fair distribution of communicative power that the NWICO proponents envisaged.

In the present era of globalization, it is not just TNCs or other institutional entities that make the planetary reach of information and communication technologies (ICTs) work for them. Citizens' and grassroots movements have formed counterglobalizing networks, using information technologies for their own objectives that emerge primarily from community needs (Keck & Sikkink 1998; Russell, 2001). Marglin (1990) held that "separating technologies from their economic, political and other freight (p. 00) enables adaptation of technology to context-specific uses without the burden of the asymmetry imposed by those who initially invented and introduced such technologies. This idea is implied in these counterglobalizing networks. Critical scholars have emphasized that technologies do not come free of history or ideology. Marglin took this observation a step further and speculated as to whether technologies could be decoupled from their political and cultural entailments. He saw an opening for such a decoupling if "deference to Western episteme" could be removed from the development picture. He argued that populist movements against the West were directed not against the technology as much as their entailments. In his NAMEDIA presentation, Pal (1983) suggested ways to achieve this decoupling by providing new historical contexts for technologies away from their places of origin. Whether the satellite communication strategies that Pal suggested in this context could engender new forms of domination is another question that needs to be addressed separately.

DEVELOPMENT, POSTDEVELOPMENT, AND GLOBALIZATION

From the perspective of developing regions in recent years, states seem to view globalization more as an engine of modernization than as a harbinger of a postmodern economy. For example, Jackson and Mosco (1999) demonstrated that Malaysia's multimedia supercorridor, a physical manifestation of postindustrial society, was in fact inspired by the modernization rhetoric of "speed[ing] development down a worldwide information highway" (p. 34). Others note a distinct shift to a new global era that redefines old problems. Global media empires now colonize the imagination by aggressively promoting the pleasures of consumption, persuading consumers to desire products manufactured by TNCs (Thussu, 1998).

Supranational organizations and their wings, such as the World Bank (WB) and the United Nations Development Programme (UNDP) of the UN, are now engaged in a struggle to integrate the international developmentalist approach with the newer mandates of globalization to address development. Their current rhetorical framework and policies are recognizable from the international development era—capital and (new communications) technology are emphasized. However, the weight of the rhetoric rests with the term global, as in the need for worldwide integration into the global economy, the central role of private capital, and the global interconnections through new information technologies. In recent years, the private sector and public sector management for the successful transitioning of several areas to the private sector have constituted a significant lending focus (World Bank Group, 2000). At the start of the new millennium, the United Nation Development Programme (2001) placed an overwhelming emphasis on information and communications technology as a primary globalizing agent for human development worldwide. During 2001, the UNDP introduced the technology achievement index (TAI) as one of the measures of human development—to "redefine . . . development strategies . . . in the network age" (UN Development Programme, 2001).

The WB's report on media ownership and penetration in 97 countries in 2002 (World Development Report, 2002) did not include data and extensive discussion on new media, such as computers and the Internet, presumably because of their very low to negligible numbers in most parts of the world. However, media ownership figures for the more traditional mass media, including newspapers, TV, and radio, demonstrate a continuing pattern from the last half century or so—more private ownership of newspapers (57%) as opposed to the dominance of state ownership of broadcast media (60% for TV and 72% for radio; United Nations Development Programme, 2001). Media penetration (one indication of media access) continues to be limited in the case of many of these 97 countries. For example, although radio sets per 100 people were the highest worldwide among the three media, 16.3 sets in Africa point to a persistent lack of access compared with the 86.4 sets in the Organization for Economic Cooperation and Development (OECD) countries. Newspapers, the medium requiring higher literacy levels compared with broadcast media, continue to have low penetration rates—under 15% for all regions except the OECD countries. Overall, media access in Africa, Asia and Pacific (non-OECD countries) and the Middle East/North Africa regions is the lowest (United Nations Development Programme, 2001).

Data on new media penetration demonstrate the acute nature of the global digital divide. These patterns of access echo the access to the traditional mass media in an earlier era. Given the general population distribution worldwide, Internet use is extensive in the United States and other high-income OECD countries, significantly lower in Latin America, the Caribbean, and East Asia/Pacific, and the lowest in the Arab states, Sub-Saharan Africa, and South Asia (the United Nation Development Programme, 2001). A new digital geography has emerged, where the world is now divided into leaders, potential leaders, dynamic adopters, and marginalized groups based on technological innovations. The global hubs of technological innovations (based on the new communications media) are concentrated most intensely in Europe and North America, with a few hubs scattered sparsely in other regions (United Nation Development Programme, 2001). Thus, the material patterns, the ideology of the international development era, and the modernization mindset continue to figure prominently in the global era at the suprastate level.

The problems created by the development discourse have not disappeared in the wave of globalization, but instead key disparities among nations created by the development discourse are exacerbated. However, evidence points to significant changes in the global era as well. The question remains as to whether globalization has splintered the hitherto cohering idea of communication and development by directing our attention to global capitalism and how the local communities affected by the spread of global capital are responding to heightened development needs at present. (see Table 5.1).

The charge now is that older theories and paradigms of development are severely challenged because they have been unable to explain how social change in many regions of the world bears little resemblance to the transformations envisaged for them by advocates of the development project. Some have begun establishing that we are now in a post-development era (Escobar, 1992, 2000; Rahnema, 1998). Increasingly, agencies such as the WB fund states to privatize economies and social programs previously constituting the development activities of the state. At a more microsocial level, communities continue to seek solutions to social, economic, infrastructural, and

TABLE 5.1. COMMUNICATION AND DEVELOPMENT IN THE INTERNATIONAL AND GLOBAL ERAS

Continuities and Changes	International Era	Global Era
Continuities		
a. International/global space	Hierarchical, based on capital and technology	Hierarchical, based on capital and technology
b. Communication technology and development	Wide disparities in information and communication technologies among regions	Further disparities in information and communication technologies among regions
c. Media access	Low in developing regions	Low in developing regions
d. Social change	Modernization paradigm/modernity top–down approach at the national level	Modernization paradigm, top–down approach continues
Changes		
a. International/global space	National boundaries more prominent—territorial	National boundaries highly contested—virtual
b. Communications media and development	Media for transmission of development	New media for creating networks to mobilize social change
	World Bank, UNDP emphasis on industrialization and building material economies	WB, UNDP emphasis on new information and communication (I&C) technologies and building knowledge economies
c. Media and culture	Politics of representation of national images	Absorption of the symbolic (design, aesthetics, identity) in the global cultural economy
d. Development funding and aid	To states for development programmes	To states to help privatize development activities and build markets
e. Social change	State as main agency effecting social change—development	Increasingly, local communities, NGOs, and new social movements define and effect social change at local levels

now ecological dilemmas, often with the aid of nongovernmental organizations (NGOs) and independent of state action. Alternative avenues are being sought in a time when traditional institutions constituting the development machinery and the top–down approach to development are losing their power as engineers of social change in developing societies (Fair & Shah, 1997; Huesca, 2003). It is important in this light to integrate analysis of globalization with analysis of the current state of communication and development. We would have to examine redefinitions of development, the role of communication in this redefinition, and the social composition and organization of entities engaged in redefining their material and symbolic needs.

CONCLUSION

It is difficult to clearly define the global turn as a radically different historical phase from the previous international development era. This is especially the case for communication scholars because they have dealt with the objects of analysis and their outcomes (i.e., information technologies, hegemonic global media cultures, etc.) across both of these eras. Yet there is a significant change in surrounding institutional power arrangements, popular response, and a fluid and negotiatory environment to warrant the treatment of the global in its present form as a new phase in the world history of the present. Such awareness and reflexivity would restrain overly optimistic claims about the potential of global capitalism to bring unprecedented economic opportunity and prosperity to the developing regions (revisiting the modernization arguments). It would also necessitate a careful consideration of the democratic potential of new communications media, with an eye toward questions of access and the requisite technological literacy, along with the limits placed by global markets.

Globalization has led to the spread of a postmodern global culture as well as the increasing emergence and visibility of local popular culture expressed through a combination of traditional, modern, and new media. Symbolic goals (such as identity) help us understand how the political project of struggle for both survival with identity and social change are recast and recodified at the local level because globalization has altered social landscapes.

6

CONCLUSION

Within the 10-year period of intense debates for a NWICO, the discourse of communication and development has defined both terms in specific, albeit limited ways, and yet has opened up alternate interpretive possibilities. These modes of understanding informed many national development plans for at least 40 years, and continue to play a decisive role in international funding for development. The first four tropes discussed in this study—enumeration, macrospeak, surveillance, and invisibility—constructed boundaries within which conceptions of communication, society, and social change could occur. They established that communication, through its many guises, was instrumental in achieving development. In this reading, communication becomes technology, economy, functional apparatus, symbol of modernization, and preserver of tradition. Third World development would eventually guarantee the participation of these countries in the constitution of an international public sphere.

The conditions of possibility for understanding communication and development in ways that do not coincide with the dominant discourse are expanded when we look for alternate definitions in the tropes of resistance. Theoretical and potentially different ways of perceiving the world order/international arena point to alternative communication, incorporation of instruments of modernity in the histories of developing countries, and an understanding of communication as a less instrumental and, perhaps more organic social phenomenon.

This study used the NWICO debates to understand the construction of a discourse of communication and development as an ideology that has emerged from, among other things, a struggle for the social construction of the meanings of communication, development, and change. As Layder (1994) said of ideology, it is a collection of "systems of ideas...employed in ways with which to attempt to justify or rationalize forms of domination and make them seem natural and eternal" (p. 106). We also saw in the last chapter that naturalization of such an idea has been challenged, thereby demonstrating the provisional unity of texts that articulated the discourse of communication and development (Allan, 1993).

The tropes presented in this study by no means exhaust the discursive positions apparent in the discourse of communication and development, identified by both proponents of and opponents to a new international communication order. The reading of a dominant discourse is also a work of the author/researcher; it occurs within a certain theoretical framework and hence, while unique in the reading in some respects, also conveys common and recognizable ideas and ideologies. In this vein, I have provided a close reading of some of the tropes in selected texts. However, it is hoped that they provide a glimpse of the dynamics of the discourse of communication and development and also point to new directions for understanding communication in relation to social change, and its constitutive cultural role in Third World societies.

The discourse is expressive of the larger ideology of communication and development. This ideology has established ways of thinking about the international space when reference was made to the achievement of a democratic world order and a global public sphere. The strategies analyzed in the study hint at the power to control and legitimate certain knowledges as universal. International organizations, funded by economically powerful nations, envisioned certain types of communicative arrangement for all societies. Support for as well as resistance to such arrangements is apparent in the NWICO policy debates, as the study has demonstrated. In official discourse in particular, the concept of development served as a major cultural reference point for planning and implementing moves that have, as several critics have observed, served to maintain inequality among nations.

The period of the NWICO debates was also rife with instances of negotiation with the dominant discourse of development. Negotiation of meanings for understanding the world order and communications through multiple inflections served to demonstrate, in Bakhtin's (1981) term, the heteroglossic nature of the term communication and development. Resistance in the official and diplomatic circles was organized around familiar meanings of development—signified by technology in the main part. No large radical change was visible as a result of such resistance or constructions of alternate meanings of phenomena, except perhaps with instances in the realm of alternate communication or the establishment of alternate regional news agencies. But the dominant order continues to be challenged. Recent and even not-so-recent attempts at more radical approaches to social change seem to

Conclusion

have eluded the net of the ideology of communication and development. How does one view these changes and initiatives at the grassroots level that seemingly place themselves beyond development? Attempts to understand these changes are apparent in institutional struggles to shift development orientation from the international to the global eras. Newer forms of doing development have emerged, and new debates about an inclusive global democracy through new media have emerged. Because of the power and potential of information and communication technologies (ICTs) today, democracy is tied to technology, which constitutes a significant portion of the development discourse.

NEW MEDIA, NEW DEVELOPMENTS, AND OLD QUESTIONS

Newer media of communication, particularly the computer, the Internet, satellite radio, and cellular telephony, have engendered new conversations about modernization, development, and democracy. The conceptual appeal of individuals' access to an immense storehouse of information with new technologies, and the ability of these information and communication technologies to allow them to participate in global conversation, imbues newer media such as cellular and satellite phones, and the Internet with the ideal democratic potential. In the larger conversation, the speculation is that developing nations would leapfrog with ICTs into development, obtaining information about business, health care, education, and a host of other areas more easily than ever before. At the same time there is the reality of the global digital divide, where a majority of people is yet to acquire basic literacy (besides technological literacy) and are far from gaining access to new technologies. The inexorable force of ICT development and the large and looming global digital divide have brought to the fore once again questions of development and democracy. These questions form the larger backdrop against which global debates about the new information society take place, such as the World Summit on the Information Society (WSIS). Familiar actors such as the International Telecommunication Union and the UNESCO now debate the information society, along with participants from the private sector and from what has been labeled in a broad and catch-all term as *civil society*.

In this context, the WSIS has become the center of debate, deliberation, and plans for action in the international/global communication community. Initiated by the International Telecommunication Union (ITU) and conducted under the auspices of the United Nations, the WSIS is charged with addressing the creation of a global information society with the help of multiple actors. Besides governmental and intergovernmental representatives, partners from the private sector and representatives of civil society at large are also involved in the construction of a global information society. This last constituency includes activist organizations, NGOs, media organizations, and a host of others that did not figure as directly or prominently in the NWICO debates. All stakeholders are expected to unite in "digital solidarity" in the attempt to create a global knowledge society (WSIS Declaration of Principles, http://www.itu.int/ wsis/docs/geneva/offiearcial/dop.html).

One of the striking aspects to the WSIS is its deep engagement with the discourse of modernization. Technology forms a cornerstone in the vision of a global information society, as does economics (the role for the private sector is clear in both the Declaration of Principles and the Plan of Action). Access to ICTs, appropriate training for using them productively, global standardization of technology and competencies, and "stability, predictability and fair competition" (Declaration of Principles) have been identified as fundamental to achieving a global knowledge society. The emphasis is on information as product and engine of technologies intended to produce "people-centred, inclusive and development-oriented Information Society" (Declaration of Principles). A robust private sector is seen as essential to achieving the goal of a global information society, and in this context, national policies creating a favorable climate for foreign direct investment have been encouraged. The Declaration emphasizes the importance of small- and medium-sized enterprises as a necessity for making ICTs work for development purposes at the more local levels. The intention is to create a global society that is "multilateral, transparent, and democratic" (Declaration of Principles).

The Plan of Action of the WSIS, also finalized at the first summit in Geneva in 2003, calls for sustainable e-strategies at the national level. Connectivity, education and training, confidence, and security are emphasized as fundamental requirements for the establishment of a successful information society. The prefix *electronic* (or *e-*) has been attached to many national domains such as e-governance, e-health, and so on, which specify areas in which states should achieve targeted and measurable goals. Both the Declaration and the Plan of Action are broadly inclusive of multiple domains, factions, and concerns. For example, private sector responsibility is emphasized and, at the same time, concentration of ownership is discouraged. Marginalized populations of every kind—including women, children, refugees ethnic minorities—are addressed in these documents. All are enfolded into the rhetoric for a future *Information Society* (I and S in the uppercase; the term is a proper noun in WSIS documents) that is global, subordinating states, regions, and various combinations of stakeholders at the international, national, regional, and local levels to this larger society.

As with any declaration at the global level involving multiple parties, the compulsion to inclusiveness comes at the cost of productive specificity. In the case of the WSIS, the need for sharper definitions of roles, actions, and outcomes has been picked up in a dialogic capacity by organized interests that advocate for accountability of the decision makers in the creation of the Information Society. A civil society consortium registered as a valid stakeholder in the WSIS discussions, Communication Rights for the Information Society (CRIS), is a "group of NGOs involved in media and communication around the world" (CRIS Charter, http://www.crisinfo.org/content/view/full/78). CRIS has created its own charter, vision, and committees, and it maintains a Web site that is updated frequently with reports drawn from various sources of interest to CRIS members and supporters. CRIS demands social and economic justice that may be interpreted as a more specific (and more activist) articulation of the declaration of ICTs for the development of and benefit for all that is more characteristic of WSIS. Also in an activist vein, the CRIS Charter has reopened a debate with the (ITU) in its goal to "to reclaim the airways and spectrum as public commons and to tax commercial use for public benefit." This is part of the larger role that CRIS envisions for all media in the Information Society, beyond the ICTs. The question of access, important to both WSIS and CRIS goals, is extended by CRIS to demand that the right be "realized in the public domain." Further, CRIS specifies access to "information societies as a whole" as an important interpretation of the fundamental right to communicate. Especially in a domain such as ICTs, where the only constant is change in technologies, keeping up-to-date (to make the technology useful and productive for sustainable development) requires such access, along with continuous training for operating changing technologies.

At this early stage in the debates and attempts to put in place plans of action, no clear picture about the outcome of the WSIS is available except informed speculation after careful observation that the civil society platform will eventually have a marginalized and largely silent role in the WSIS, especially because the platform is not vested with decision-making powers. Its status as lobbyist, negotiator between the Summit Secretariat and various stakeholders in the WSIS, and its advisory role with governments have kept the civil society platform under the control of the ITU, the UN, national governments, and also the private sector. CRIS reports on WSIS issues present a gamut of uncertainties and emerging problems faced by all participants of the WSIS (civil society platforms included). The following are a few examples. The accreditation criteria and, as Raboy (2005); pointed out, the participation process for civil society organizations in advising intergovernmental organizations and governments have turned the civil society consortium into a bureaucratic entity with factions within this platform. Concerns about issues of data storage and the resultant continuation of the existing power/knowledge nexus have resurfaced in the Internet Age in the discussion of stocktaking in and by the ITU, similar to the concerns over the ownership of data derived from satellite remote sensing in the 1980s (as discussed in the trope of surveillance in an earlier chapter). These concerns are punctuated by the hope that advisory groups, especially the Working Group on Internet Governance (WGIG), are making some progress in bringing critical issues for decision making to the attention of the Summit Secretariat (Bendrath, 2005). The WGIG's effort to expand the diversity in board membership for Internet governance, to ensure decision-making capacities for more than the private sector or members from select developed countries, is an example.

The WSIS and the PrepComs (Preparatory Committee meetings that include intergovernmental, governmental, private sector, and civil society representatives) preceding Phases 1 and 2 of the WSIS demonstrate interesting similarities and differences in comparison with the NWICO. The latter was a policy effort intended to establish a new world communication order, and WSIS, a forum for articulating a new communication order for a new world in the information age. In both instances, plans of action have been formulated, and both the NWICO and the WSIS draw on the same vocabulary to express visions of a new world order based on information/communication and inclusiveness. Similarly, both efforts at achieving global democracy in communication foreground the right to communicate as central to achieving a new world order and draw on the Universal Declaration of Rights as a founding text.

Several areas of concern pertaining to media technologies, information flow, and questions around access to communication facilities emerged from the NWICO debates. Technology, democracy, and development underpin both the NWICO and WSIS debates. Assertions of the idea of distributive justice in communications power, whatever the side and constituency may be, are also apparent in both the debates. We can understand both the NWICO and noncommercial Civil Society platforms of the WSIS as organized voices striving to articulate and establish alternate global democracy, recognizing related problems and issues in developing regions. The Civil Society component, as a principal stakeholder, is especially active in critical and dialogic responses to intergovernmental, national, and private sector proposals and decisions to chart a path to an Information Society. CRIS has positioned itself as one such entity. Other concerns now under discussion include proposals to fund community initiatives to integrate the local into the global information society that are being treated with caution and skepticism by the other stakeholders.

However, differences between the NWICO and the WSIS-led ideas for a new world communication order are also apparent. In the context of the NWICO, members of developing nations and their supporters constituted

Conclusion

the primary proponents of the policy idea that would ensure a new world communication order. In the case of the WSIS, the base is expanded to include the private sector and civil society platforms. For example, there are differences in legitimizing procedures and the composition of concerned parties typically seen as the Other to a dominant position in the NWICO and WSIS. The composition of noncommercial Civil Society platforms of the WSIS ranges more broadly compared with the governmental and expert members participating in the NWICO debates. Noncommercial civil society members originate from multiple cause-driven backgrounds and advocate a range of issues from the environment to antiwar to the right to communicate. There is also a shift in the media under consideration—from a broad range of media, including traditional mass media to satellite communication, in the NWICO to digital media—in the WSIS.

The NWICO research teams were appointed/organized from within the UNESCO and related (mostly) supranational organizations, whereas the private sector and Civil Society constituencies of the WSIS (both for-profit and nonprofit) were organizationally and bureaucratically integrated from without. These observations might leave us with a few important questions that return to issues of power, dominance, development, and democracy. Does this broader base in the Civil Society group indicate more power, especially for the noncommercial mobilizations that constitute an other to regulatory and policy organizations such as the International Telecommunication Union or the Internet Corporation for Assigned Names and Numbers (ICANN)? Where do grassroots virtual communities figure in the WSIS debates in ways grassroots collectives could not in the NWICO debates? Where exactly might the discursive contestation between the WSIS and the noncommercial Civil Society constituents lie? Where do the WSIS arguments reflect a continuity from the NWICO debates, and where do they diverge?

Other innovative models of development have emerged in recent years. One noteworthy model—capitalist development—marries global capitalism and information technology with rural development. In recent years, for example, Greenstar, a virtual organization that characterizes itself as a development corporation, has built a base of solar energy projects in rural areas worldwide that would eventually be sustained by e-commerce. The rhetoric of this development corporation is empowerment, and it argues are for debt-free development—a potent attraction for many developing regions perpetually in the grip of development debts. The intention is to eventually cede control to communities, using commission money from the sale of digital culture products through e-commerce for new projects, to preserve local cultures digitally, and share the rich traditions these communities have to offer by making them available worldwide on the Internet (Sosale, in press).

Radhakrishnan's (1996) understanding of postcolonial theorist Gayatri Spivak's sentiments about the fragmentation of communities hitherto perceived as binding and cohesive—that might result from postdevelopment approaches such as grassroots movements, e-commerce communities, new networked societies, and so on—helps to bring to the fore the political project in such pluralities. Rather than just accepting and celebrating differences, a persistent and sustained critique of the existing orders is necessary to dispel any sense of complacency that might derive from the assumption that such communities will co-exist peacefully and harmoniously. In this sense, the understandings of communication and development would be constantly and critically evaluated within defined social contexts.

Several factors point to this study as only a beginning—for example, the brief time period selected for the study, the limited number of texts and tropes analyzed, the breadth of cultures and media that still need to be considered, and the brief treatment of media professionalism. Examining the phenomenon within a limited time frame is only a beginning to help understand the construction of the discourse of communication and development.

Selected communicative practices within the discourse of communication and development also require further research. For example, even within the 10-year period of the NWICO debates, considerable discussion was generated about journalistic ways of seeing and knowing. Frameworks offered by critical media scholars such as Fishman (1988), Gitlin (1980), Schudson (1995), and Tuchman (1978) in the American context could serve as springboards for studying the discourse of journalistic practices and professionalism in the NWICO debates. Shah's (1996) work on the development of a humanistic journalism is both a valuable start and a sign for taking critical media research further into the domain of development journalism.

Participant observation and ethnographic studies such as Huesca's in Bolivia, besides similar studies in Third World pockets in metropolitan centers— such as Michaels' study of aboriginal television in Australia (see Chanter, 1991, for discussion) and a study of media and a housing policy movement in Finland (Saukko, 1996)—point to considerable enhancement of understanding the discourse of communication and development in the developing world in the present time, by no means completely dissociated from the concerns expressed in the NWICO policy debates. The new avenues opened up by the WSIS offer a wealth of possibilities for future study. Much remains to be done, and this attempt at mapping this discourse is a beginning. Although a beginning leaves many things unexplored, it also provides opportunities for further research, and I hope that this study has provided some in the domain of communication, development, and social change.

REFERENCES

Addo, H. (1984). Introduction: Pertinent questions about the NIEO. In H. Addo (Ed.), *Nine critical essays on the New International Economic Order* (pp 1–17). Tokyo: United Nations University.

Allan, S. (1993). *No truth, no apocalypse: Dismantling the cultural dynamics of nuclearism in official and news discourse.* Unpublished doctoral dissertation, Carleton University, Toronto, Canada.

Alvares, C. (1992). Science. In W. Sachs (Ed.), *The development dictionary: A guide to knowledge as power* (pp. 219-222). London: Zed.

Appadurai, A. (1990). Technology and the reproduction of values in rural western India. In F. A. Marglin & S.A. Marglin (Eds.), *Dominating knowledge: Development, culture, and resistance* (pp. 185–216). Oxford: Clarendon.

Appadurai, A. (1993). Disjuncture and difference in the global cultural economy. *Public Culture, 2*(2), 1–24.

Appadurai, A., Korom, F. J., & Mills, M.A. (1991). *Gender, genre, and power in South Asian expressive traditions.* Philadelphia: University of Pennsylvania Press.

Apter, D. (1987). *Rethinking development: Modernization, dependency, and postwar politics.* Newbury Park, CA: Sage.

Atton, C. (2003). *Alternative media.* London, Thousand Oaks, New Delhi: Sage.

Babe, R. (1995). *Communication and the transformation of economics: Essays in information, public policy, and political economy.* Boulder, CO: Westview.

Bakhtin, M. (1981). *The dialogic imagination: Four essays* (M. Holquist, ed., C. Emerson & M. Holquist, trans.). Austin: University of Texas Press.

Bello, W. (1996). Fast-track capitalism, geoeconomic competition and the sustainable development challenge in East Asia. In C. Thomas & P. Wilkin (Eds.), *Globalization and the South* (pp. 143-162). New York: St. Martin's Press.

Bendrath, R. (2003). Civil Society and the multi-stakeholderism: Discussion emerging about opportunities and strange bedfellows. http://www.crisinfo.org/newsroom/wsis_phase_2_special_prepcoms/civil_society_and_the_multi_stakeholderism_discussion_emerging_about_opportunities_and_strange_bedfellows (accessed 05/ 21/ 06).

Berger, M. (2001). The rise and demise of national development and the origins of post-Cold War capitalism. *Millennium, 30*(2), 211-234.

Berrigan, F. (1981). Community communications: The role of community media in development. *Reports and Papers on Mass Communication, 90* (49 p). Paris: UNESCO.

Bielenstein, D. (1979). Findings and recommendations of the international conference 'Toward a New World Information Order: Consequences for Development Policy.' In *Toward a new world information order: Consequences for Development Policy: An international conference* (pp. 87-114). Bonn, FRG: Institute for International Relations and the Freidrich-Ebert-Stiftung.

Bloch, M. (1975). Introduction. In M. Block (Ed.), *Political language and oratory in traditional society*. London, New York, San Francisco: Academic Press.

Bourdieu, P. (1991). *Language and symbolic power* (J. B. Thompson, Ed.; G. Raymond & M. Adamson, Trans.). Cambridge: Harvard University Press.

Bourdieu, P. (1994). Structures, *habitus* and practices. In *The polity reader in social theory*. Cambridge: Polity Press.

Boyd-Barrett, O. (1980). *The international news agencies*. London: Constable.

Boyd-Barrett, O. & Thussu, D. (1993). NWICO strategies and media imperialism: The case of regional news exchange. In K. Nordenstreng & H. Schiller (Eds.), *Beyond national sovereignty: International communication in the 1990s* (pp. 177-192). Norwood, NJ: Ablex.

Boyd-Barrett, D., & Thussu, D. K. (1993). *Contra-flow in global news:International and regional news exchange mechanisms*. London: J. Libbey.

Braman, S. (1990). Trade and information policy. *Media, Culture and Society, 12*, 361–385.

Burton, F., & Carlen, P. (1979). *Official discourse: On discourse analysis, government publications, ideology and the state*. London, Boston, and Henley: Routledge & Kegan Paul.

Canclini, N.G. (1988). Culture and power: The state of research. *Media, Culture & Society, 10*, 467-497.

Chakravartty, P. (2001). Flexible citizens and the internet: The global politics of local high-tech development in India. *Emergences, 11*(1), 69-88.

Chanter, A. 1991. The redevelopment of aboriginal communications. In F. L. Casmir (Ed.), *Communication and development* (pp. 250-269). Norwood, NJ: Ablex.

Chatterjee, P. (1993). *Nationalist thought and the colonial world: A derivative discourse*. Minneapolis: University of Minnesota Press.

Chouliaraki, L. & Fairclough, N. (1999). *Discourses in late modernity*. Edinburgh: Edinburgh University Press.

Cleevely, D., & Walsham, G. (1980). Telecommunications models: Planning for regional development in LDCs. *Telecommunications Policy, 4*(2), 108-118.

Contreras, E. (1980). Brazil and Guatemala: Communications, rural modernity, and structural constraints. In E. McAnany (Ed.), *Communication in the rural Third World: The role of information in development* (107-145). New York: Praeger.

Crush, J. (1995). *Power of development*. London, New York: Routledge.

References

Derrida, J. (1986). Différance. In M. C. Taylor (Ed.), *Deconstruction in context: Literature and philosophy* (pp. 396–420). Chicago: The University of Chicago Press.

Development Dialogue. (1981). Statement by the participants in the 1975 Dag Hammarskjöld Third World journalists' seminar. *Development Dialogue, 2*, 115-118.

Dews, P. (1986). Editor's introduction. In *Habermas: Autonomy and solidarity: Interviews* (pp. 1–34). London: Verso.

Doty, R. (1996). *Imperial encounters: The politics of representation in North–South relations*. Minneapolis: University of Minnesota Press.

Downing, J. (2001). *Radical media: Rebellious communication and social movements*. Thousand Oaks, CA: Sage.

The Drishti Media Group. (1995). http://home.dti.net/foil/ resources/drishti.htm

Escobar, A. (1992). Planning. In W. Sachs (Ed.), *The development dictionary: A guide to knowledge as power* (pp. 132–145). London: Zed.

Escobar, A. (1995). *Encountering development: The making and unmaking of the Third World*. Princeton, NJ: Princeton University Press.

Escobar, A. (1999). Discourse and power in development: Michel Foucault and the relevance of his work to the Third World. In J. Servaes & T. Jacobson (Eds.), *Theoretical approaches to participatory communication* (pp. 309-336). Cresskill, NJ: Hampton Press.

Escobar, A. (2000). Place, power, and networks in globalization and postdevelopment. In K. Wilkins & B. Mody (Eds.), *Redeveloping communication for social change: Theory, practice, and power* (pp. 163-173). Lanham, MD: Rowman and Littlefield.

Esteva, G. (1992). Development. In W. Sachs (Ed.), *The development dictionary: A guide to knowledge as power* (pp. 6–25). London: Zed Books.

Esteva, G. (1996). Hosting the otherness of the other. In F. Appfel-Marglin & S. A. Marglin (Eds.), *Decolonizing knowledge: From development to dialogue*. Oxford: Clarendon Press; New York: Oxford University Press.

Fair, J. E. (1996). The body politic, the bodies of women, and the politics of famine in US television coverage of famine in the Horn of Africa. *Journalism and Mass Communication Monographs, 158*.

Fair, J.E., & Shah, H. (1997). Continuities and discontinuities in communication and development research since 1958. *Journal of International Communication, 4*, 3-23.

Featherstone, M. (1990). Global culture: An introduction. In M. Featherstone (Ed.), *Global culture: Nationalism, globalization and modernity* (pp. 1–14). London: Sage.

Fishman, M. (1988). *Manufacturing the news*. Austin: University of Texas Press.

Fiske, J. (1993). Local knowledges. In *Power plays power works* (pp. 181–205). London: Verso.

Foucault, M. (1972). *The archaeology of knowledge and the discourse on language*. New York: Pantheon.

Foucault, M. (1978). *The history of sexuality, Volume 1*. New York: Pantheon.

Foucault, M. (1980). *Power/knowledge: Selected interviews and other writings 1972–1977* (C. Gordon, Ed.; C. Gordon, L. Marshall, J. Mepham, & K. Soper, Trans.). New York: Pantheon.

Galtung, J. (1982). The New International Order: Economics and communication. In M. Jussawalla & D. Lamberton (Eds.), *Communication economics and development* (pp. 133-143). New York: Pergamon Press.

Gauhar, A. (1981). Third World: An alternative press. *Journal of International Affairs, 35*(2), 165-177.

Gazin, R. 1978. *Communication technologies of the 1980s: Recent progress and its impact upon communication policy and development. International Commission for the study of communication problems, 82* (pp. 13-19). Paris: UNESCO.

Germain, R. (2000). Globalization in historical perspective. In R. Germain (Ed.), *Globalization and its critics: Perspectives from political economy* (pp. 67-90). New York: St. Martin's Press.

Giddens, A. (1993). *New rules of sociological method: A positive critique of interpretative sociologies* (2nd ed.). Stanford, CA: Stanford University Press.

Gitlin, T. (1980). *The whole world is watching: Mass media in the making and the unmaking of the new left*. Berkeley, Los Angeles, London: University of California Press.

Golding, P. (1974). Media role in national development: Critique of a theoretical orthodoxy. *Journal of Communication, 24*, 39-53.

Golding, P. (1978). Media professionalism in the Third World: The transfer of an ideology. In J. Curran, J. Woollacott, & M. Gurevitch (Eds.), *Mass communication and society* (pp. 291-308). London: Edwin Arnold.

Gupta, A. (1998). *Postcolonial developments: Agriculture in the making of modern India*. Princeton, NJ: Princeton University Press.

Hall, S. (1980). Encoding/decoding. In Centre for Contemporary Cultural Studies (Ed.), *Culture, media, language: Working papers in cultural studies 1972–79* (pp. 128–183). London: Hutchinson.

Hall, S. (1985). The rediscovery of "Ideology": Return of the repressed in med studies. In V. Beechey & J. Donald (Eds.), *Subjectivity and social relations: A reader*. Milton Keynes, PA: Open University Press.

Hall, S. (1997). Representation and the media [Video]. Northampton, MA: Media Education Foundation.

Hamelink, C. (1984). *Transnational data flows in the information age*. Lund: Studentlitteratur AB, Chartwell-Bratt.

Hardt, M., & Negri, A. (2000). *Empire*. Cambridge, MA: Harvard University Press.

Hornik, R. (1988). *Development communication: Information, agriculture, and nutrition in the Third World*. New York: Longman.

Huesca, R. (1994). A procedural view of participatory communication: Lessons from Bolivian tin miners' radio. *Media, Culture & Society, 17*, 101–119.

Huesca, R. (2003). Tracing the history of participatory approaches to development: A critical appraisal. In J. Servaes (Ed.), *Approaches to development: Studies on communication and development* (pp. 8-1-8-36). Paris: UNESCO.

Huesca, R., & Dervin, B., (1994). Theory and practice in Latin American alternative communication research. *Journal of Communication, 44*(4), 53–73.

Hussein, S. 1980. Main forms of traditional communication: Egypt. *International Commission for the Study of Communication Problems, 93*. Paris: UNESCO.

Jackson, S., & Mosco, V (1999). The political economy of the new technological spaces: Malaysia's multimedia super corridor. *Journal of International Communication*, 6(1), 23–40.

Jacobson, T. (1996). Development communication theory in the 'wake' of positivism. In J. Servaes, T.L. Jacobson, & S.A. White (Eds.), *Participatory communication for social change*. New Delhi, Thousand Oaks, London: Sage.

Jacobson, T., & S. Kollurii, S. (1999). Participatory communication as communicative action. In J. Servaes & T. Jacobson, (Eds.), *Theoretical approaches to participatory communication* (pp. 265–280). Cresskill, NJ: Hampton Press.

Jussawalla, M., & Lamberton, D. M. (1982). Communication economics and development: An economics of information perspective. In M. Jussawalla & D. M. Lamberton (Eds.), *Communication economics and development* (pp. 1-15). New York: Pergamon Press, with The East West Center, Hawaii.

Kang, J. M. (1988). *The politics of the New World Information and Communication Order: A historical analysis of the conflict between the Third World and the United States, 1970–87*. Unpublished doctoral dissertation. Madison: University of Wisconsin–Madison.

Keck, M., & Sikkink, K. (1998). *Activists beyond borders*. Ithaca, NY: Cornell University Press.

Kidd, R. 1983. Popular theater and popular struggle in Kenya: The story of the Kamiriithu Community Educational and Cultural Centre. *IFDA Dossier, 33*, 17–29.

Komatsuzaki, S. (1978). Communication technologies of the 1980s: The social implications. *International Commission for the Study of Communication Problems, 83*. Paris: UNESCO.

Kothari, S., & Sen, J. (1983). Communicating and development. *IFDA Dossier, 33*, 3–16.

Laclau, E., & Mouffe, C. (1985). *Hegemony and socialist strategy: Towards a radical democratic politics*. London: Verso.

Lash, S., & Urry, J. (Eds.). (1994). *Economies of signs and space*. London: Sage.

Layder, D. (1994). *Understanding social theory*. London, Newbury Park, New Delhi: Sage.

Lee, C.C. (1980). *Media imperialism reconsidered*. Beverly Hills: Sage.

Lent, J. (1995). *Asian popular culture*. Boulder, CO: Westview.

Lent, J., and Vilanilam, J. (1979). *The use of development news: Case studies of India, Malaysia, Ghana and Thailand*. Singapore: Asian Mass Communication Research and Information Centre.

Lerner, D. (1956). *The passing of traditional society: Modernizing the Middle East*. Glencoe, IL: The Free Press.

Lummis, D. (1992). Equality. In W. Sachs (Ed.), *The Development Dictionary: A guide to knowledge as power* (pp. 38–52). London: Zed Books.

Malik, M. (1983). Traditional forms of communication and the mass media in India. *Communication and Society, 13* (101 pp). Paris: UNESCO.

Marglin, S. (1990). Towards the decolonization of the mind. In F. A. Marglin & S. A. Marglin (Eds.), *Dominating knowledge: Development, culture, resistance*. Oxford: Clarendon.

Martin, G., & Hindley, P. (1981). Communication at an impasse: A plea for humanized communication. *Development Dialogue, 2*, 105-112.

Martín-Barbero, J. (1988). Communication from culture: The crisis of the national and the emergence of the popular. *Media, Culture & Society, 10*, 447–465.

Martín-Barbero, J. (1993). *Communication, culture and hegemony: From the media to mediations* (F. Fox & R.A. White, Trans.). London, Newbury Park, New Delhi: Sage.

Martinussen, J. (1997). *Society, state and market: A guide to competing theories of development*. London: Zed Books.

Matta, F.R. (1979). Transnational communication and the alternative communication answer. *Current Research on Peace and Violence, 2*(2), 96-103.

Mattelart, A. (1994). *Mapping world communications: War, progress, culture* (S. Emmanuel & J. A. Cohen, Trans.). Minneapolis: University of Minnesota Press.

McAnany, E., & Oliveira, J. M. (1980). The SACI/ EXERN project in Brazil: An analytical case study. *Reports and Papers on Mass Communication, 89* (46 pp.). UNESCO/ MC/ CH/89. Paris: UNESCO.

Melucci, A. (1992). Liberation or meaning? Social movements, culture and democracy. *Development and Change, 23*(3), 43-77.

Mohanty, C. T. (1991). Under Western eyes. Feminist scholarship and colonial discourses. In C. T. Mohanty, A. Russo, & L. Torres (Eds.), *Third World women and the politics of feminism*. Bloomington: Indiana University Press.

NAMEDIA. (1983). Final report and documents of the Media Conference of the Non-aligned. New Delhi, December 9-12. New Delhi: NAMEDIA.

Nandy, A. (1995). *The savage Freud and other essays on possible and retrievable selves*. Princeton, NJ. Princeton University Press.

Narula, U., & Pearce, B. (1986). *Development as communication: A perspective on India*. Carbondale: Southern Illinois University Press.

Nawaz, S. 1983. The mass media and development in Pakistan. *Asian Survey, 23*(8), 934-957.

Ogan, C., & Fair, J.E. (1984). A little good news: The treatment of development news in selected Third World newspapers. *Gazette, 33*, 173–91.

O'Sullivan-Ryan, J., & Kaplún, M. (1980). Communication methods to promote grassroots participation: A summary of research findings from Latin America. *Communication and Society, 6*. Paris: UNESCO.

Pal, Y. (1983). Configuring communication technologies for creative interaction within and between non-aligned and developing countries. In NAMEDIA: Final report and documents of the Media Conference of the Non-aligned (pp. 67-71). New Delhi: NAMEDIA.

Parthasarathi, G. 1983. Human and cultural values in communication. In NAMEDIA: Final report and documents of the Media Conference of the Non-aligned (pp. 54-57). New Delhi: NAMEDIA.

Pavlic, B., & Hamelink, C. (1985). *The new international economic order: Links between economics and communication*. Paris: UNESCO.

Pelton, J. 1983. The communication satellite revolution: Revolutionary change agent. *Columbia Journal of World Business, 18*(1), 77–84.

Pieterse, J. N. (1992). *Emancipations modern and postmodern*. London, Newbury Park, New Delhi: Sage.

References

Pool, I. (1990). *Technologies without boundaries: On telecommunications in a global age*. Cambridge, MA: Harvard University Press.
Poster, M. (1989). *Critical theory and poststructuralism: In search of a context*. Ithaca, NY: Cornell University Press.
Raboy, M. (2005). The WSIS prepares to grapple with Internet governance. http://www.crisinfo.org/content/view/full/799 (accessed 06/13/05).
Radhakrishnan, R. (1996). *Diasporic mediations: Between home and location*. Minneapolis: University of Minnesota Press.
Rahnema, M. (1992). Participation. In W. Sachs (Ed.), *The development dictionary: A guide to knowledge as power* (pp. 116–131). London: Zed Books.
Rahnema, M. (1998). *The post-development reader*. London: Zed Books.
Randall, V., & Theobald, R. (1985). *Political change and underdevelopment: A critical introduction to Third World politics*. Durham, NC: Duke University Press.
Ranganath, H. 1979. Not a thing of the past. International Commission for the Study of Communication Problems, 92 (10 pp). Paris: UNESCO.
Reeves, G. (1993). *Communications and the "Third World."* London: Routledge.
Rist, G. (1997). *The history of development: From Western origins to global faith* (P. Camiller, Trans.). London: Zed Books.
Robertson, R. (1990). Mapping the global condition: Globalization as the central concept. In M. Featherstone (Ed.), *Global culture: Nationalism, globalization and modernity* (pp. 15–30). London, Newbury Park, New Delhi: Sage.
Rodriguez, C. (2001a). *Fissures in the mediascape*. Cresskill, NJ: Hampton Press.
Rodriguez, C., (2001b). Shattering butterflies and amazons: Symbolic constructions of Colombian women in development perspective. *Communication Theory, 11*(4), 472–494.
Russell, A. (2001). Chiapas and the new news: Internet and newspaper coverage of a broken cease-fire. *Journalism, 2*(2), 197–220.
Sachs, W. (1992). Introduction. In W. Sachs (Ed.), In *The development dictionary: A guide to knowledge as power* (pp. 1–5). London: Zed Books.
Said, E. (1978). The problem of textuality: Two exemplary positions. *Critical Inquiry, 4*(4), 673-714.
Saukko, P. (1996). A little village in a big world: Young squatters and the limits of news. In S. Braman & A. Sreberny-Mohammadi (Eds.), *Globalization, communication, and transnational civil society* (pp. 243-258). Cresskill, NJ: Hampton Press.
Saunders, K. (2002). Introduction. In K. Saunders (Ed.), *Feminist post-development thought* (pp. 1-38). London: Zed Books.
Savio, R. (1982). Communications and development in the 80s. *IFDA Dossier, 32*, 75-79.
Schiller, H. (1979). Transnational media and national development. In H. Schiller & K. Nordenstreng (Eds.), *National sovereignty and international communication*. Norwood, NJ: Ablex.
Schiller, H. (1989). *Culture, Inc. The corporate takeover of public expression*. New York, Oxford: Oxford University Press.
Schudson, M. (1995). *The power of news*. Cambridge, MA: Harvard University Press.

Sen, M. (1983). Influence of cinema on the people in the developing and non-aligned countries. In NAMEDIA: Final report and documents of the Media Conference of the Non-aligned (pp. 9-12). New Delhi: NAMEDIA.

Servaes, J. (1998). Human rights, participatory communication, and cultural freedom in a global perspective. *Journal of International Communication, 5*(1-2), 122–133.

Shah, H. (1996). Modernization, marginalization, and emancipation: Toward a normative model of journalism and national development. *Communication Theory, 6,* 143-156.

Shah, H. (1997). Continuities and discontinuities in communication and development research since 1958. *Journal of International Communication, 4*(2), 3–23.

Shiva, V. (2000). The world on edge. In W. Hutton & A. Giddens (Eds.), *Global capitalism* (pp. 112–129). New York: The New Press.

Shiva, V. (1997, October 9) *Sustainability in a period of globalization.* MacArthur Distinguished Lecture, University of Minnesota, Minneapolis.

Shohat, E., & Stam, R. (1994). *Unthinking Eurocentrism: Multiculturalism and the media.* London: Routledge.

Singham, A.W. & Hune, S. (1987). From Third World non-alignment to European dealignment to global realignment. In M. Kaldor & R. Falk (Eds.), *Dealignment: A new foreign policy perspective* (pp. 185-204). New York: Basil Blackwell.

Slater, D. (1992). Theories of development and politics of the postmodern—exploring a border zone. *Development and Change, 23*(3), 283–319.

Sosale, S. (2002). Communication and development in the international and global eras: Continuities and changes. *Journal of International Communication, 8*(2), 8–25.

Spivak, G. C. (1996). Bonding in difference: Interview with Alfred Arteaga. In D. Landry & G. MacLean (Eds,), *The Spivak reader: Selected works of Gayatri Chakravorty Spivak* (pp. 15–28). New York: Routledge.

Spurr, D. (1993). *The rhetoric of empire: Colonial discourse in journalism, travel writing, and imperial administration.* Durham, NC: Duke University Press.

Sreberny-Mohammadi, A. (1996). Introduction. In S. Braman & A Sreberny-Mohammadi (Eds.), *Globalization, communication and transnational civil society* (pp. 1–19). Cresskill, NJ: Hampton Press.

Steeves, L. (2001). Liberation, feminism, and development communication. *Communication Theory, 11*(4), 397–414.

Stevenson, R. (1988). *Communication, development, and the Third World.* New York: Longman.

Tehranian, M. (1978). Communication and international development: Some theoretical considerations. *International Commission for the Study of Communication Problems, 41.* Paris: UNESCO.

Tehranian, M. (1994). Communication and development. In D. Crowley & D. Mitchell (Eds.), *Communication theory today* (pp. 273-306). Stanford, CA: Stanford University Press.

Tehranian, M. (1999). *Global communication and world politics: Domination, development, and discourse.* Boulder: Lynne Reinner.

References

Thomas, C. (1996). Globalization and the South. In C. Thomas & P. Wilkin (Eds.), *Globalization and the south* (pp. 1–17). New York: St. Martin's.

Thussu, D. (1998). *Electronic empires: Global media and local resistance*. London: Arnold.

Tipps, D. (1973). Modernization theory and the comparative study of societies: A critical perspective. *Comparative Studies in Society and History, 15*, 199–226.

Tomlinson, J. (1996). Global experience as a consequence of modernity. In S. Braman & A. Sreberny-Mohammadi (Eds.), *Globalization, communication and transnational civil society* (pp. 63–87). Cresskill, NJ: Hampton Press.

Tuchman, G. (1978). *Making news: A study in the construction of reality*. New York: The Free Press.

Tunstall, J. (1977). *The media are American: Anglo-American media in the world*. London: Constable.

UNESCO. 1979. *Communication indicators 1*. CC-79/ WS/ 134 (85 pp). Paris: UNESCO.

UNESCO. (1980). *Many voices, one world: Communication and society, today and tomorrow: Towards a new more just and more efficient world information and communication order* (Final report of the International Commission for the Study of Communication Problems). Paris: Author.

United Nations. (1979). *Cooperation and assistance in the application and improvement of national information and mass communication systems for social progress and development*. A/ 34/ 148. 82-83. Annex 1-3. New York: Author.

United Nations. (1981). *Report of the United Nations Regional Seminar on Remote Sensing Applications and Satellite Communications for Education and Development* (Buenos Aires). A/AC. 105/290. (17 pp). Annex I–III. New York: Author.

United Nations. (1981). *Report of the United Nations/Economic Commission for Africa Regional Seminar on Remote Sensing Applications and Satellite Communications for Education and Development* (Addis Ababa). A/AC. 105/290. (29 pp). Annex I–V. New York: Author.

United Nations Development Programme. (2001). *Human Development Report*. Accessed June 2002, from www.undp.org/hdr2001/chapterwo.pdf.

United Nations General Assembly. (1979). *Questions relating to information: Resolution Adopted by the General Assembly at the 107th plenary meeting*. A/RES/34/181 in A/34/46, 34th session, Supplement 46. 83-85. New York: Author.

Venturelli, S. (1996). Freedom and its mystification: The political thought of public space. In S. Braman & A. Sreberny-Mohammadi (Eds.), *Globalization, communication and transnational civil society* (pp. 105–140). Cresskill, NJ: Hampton Press.

Verghese, G. 1978. Readings on the relationship between development and communication, 1. A philosophy for development communications: The view from India. *International Commission for the Study of Communication Problems, 44* (7 p). Paris: UNESCO.

Vincent, R. (1997). The new world information and communication order (NWICO) in the context of the information superhighway. In M. Bailie & D. Winseck (Eds.), *Democratizing communication? Comparative perspectives on information and power* (pp. 377–406). Cresskill, NJ: Hampton Press.

Vincent, R., & Traber, M. (1999). *Towards equity in global communication*. Cresskill, NJ: Hampton Press.

Volosinov, V. N. (1984). *Marxism and the philosophy of language* (L. Matajka & I. R. Titunik, Trans.). Cambridge, MA: Harvard University Press.

Waters, M. (1995). *Globalization*. London: Routledge.

White, H. (1978). *Tropics of discourse: Essays in cultural criticism*. Baltimore, MD: Johns Hopkins University Press.

When Women Unite: The Story of an Uprising. (1996). Documentary video screened at the 25th annual South Asia Conference, University of Wisconsin-Madison, Madison, WI.

Williamson, H. A. (1991). The Fogo process: Development support communication in Canada and the developing world. In F. L. Casmir (Ed.), *Communication and development* (pp. 270–287). Norwood, NJ: Ablex.

World Bank. (2002). *World development report*. New York: Oxford University Press.

World Bank Group. (2000). *Annual report*; Accessed June 2002, from www.worldbank.org/html/extpb/annrep2000/ibrd.htm.

WSIS Declaration of Principles. (2005). Accessed June 8, 2005, from http://www.itu.int/wsis/docs/geneva/offiearcial/dop.html

Young, R. (1981). Introduction to Michel Foucault, the order of discourse. In R. Young (Ed.), *Untying the text: A post-structuralist reader* (pp. 48–51). Boston: Routledge & Kegan Paul.

Zizek, S. (1993). *Tarrying with the negative: Kant, Hegel, and the critique of ideology*. Durham, NC: Duke University Press.

AUTHOR INDEX

A

Addo, H., 5, *121*
Allan, S., 114, *121*
Alvares, C., 24, *121*
Appadurai, A., 30(*n*5), 73(*n*2), 85, 86, 96, *121*
Apter, D., 95, *121*
Atton, C., 90, *121*

B

Babe, R., 56, *121*
Bakhtin, M., 114, *121*
Bello, W., 98, *121*
Bendrath, R., 118, *121*
Berger, M., 2, *122*
Berrigan, F., 44, 57, 62, 63, *122*
Bielenstein, D., 35(*n*7), 36, 80, *122*
Bloch, M., 27, 27(*n*4), 38, 39, *122*
Bourdieu, P., 15, 23, 24, *122*
Boyd-Barrett, D., 8, 69, 76, *122*
Boyd-Barrett, O., 8, 69, 76, 76(*n*3-4), 77, 96, *122*
Braman, S., 7, 96, *122*
Burton, F., 23, 39, *122*

C

Canclini, N.G., 102, *122*
Carlen, P., 23, 39, *122*
Chakravartty, P., 14, *122*
Chanter, A., 100, 120, *122*
Chatterjee, P., 90(*n*9), *122*
Chouliaraki, L., 22, *122*
Cleevely, D., 29, *122*
Contreras, E., 59, 60, *122*

D

Derrida, J., 11, *123*
Development Dialogue, 76, 77, 79, 84, 100, *123*
Dervin, B., 10, 59, 93, 101, 102, 103, 112, 120, *124*
Dews, P., 46, *123*
Doty, R., 10, 61, 64, 66, *123*
Downing, J., 90, *123*
Drishti Media Group, 107, *123*

E

Escobar, A., 7, 10, 48, 49, 50, 60, 61, 62, 66, 67, 91, 95, 99, 100, 101, 110, *123*
Esteva, G., 99, 108, *123*

F

Fair, J.E., 2, 4, 14, 24, 36, 37, 41, 69, 75, 80, 108, 112, 116, *123*, *126*
Fairclough, N., 22, *122*
Featherstone, M., 71, 72, 73, *123*
Fishman, M., 120, *123*
Fiske, J., 103, *123*
Foucault, M., 9, 10, 11, 12, 37, 46, 48, 49, *123*

G

Galtung, J., 78, *124*

Gauhar, A., 76, 76(n3), 80, 81, *124*
Gazin, R., 28, 29, 40, *124*
Germain, R., 74, *124*
Giddens, A., 71, 73, 74, 98, *124*
Gitlin, T., 120, *124*
Golding, P., 81, *124*
Gupta, A., 13, *124*

H

Hall, S., 9, 71, *124*
Hamelink, C., 5(n5), 13, 41, 96, *124*
Hardt, M., 14, *124*
Hindley, P., 88, 89, *126*
Hornik, R., 55, *124*
Huesca, R., 10, 59, 93, 101, 102, 103, 112, 120, *124*
Hune, S., 72, 72(n1), *128*
Hussein, S., 32, 53, *124*

J

Jackson, S., 14, 109, *125*
Jacobson, T., 10, 11, *125*
Jussawalla, M., 57, *125*

K

Kang, J.M., 22, 40, *125*
Kaplún, M., 90(n8), *126*
Keck, M., 108, *125*
Kidd, R., 104, 105, *125*
Kolluri, S., 10, 11, *125*
Komatsuzaki, S., 89, *125*
Korom, F. J., 30(n5), 73, 85, 86, 96, *121*
Kothari, S., 101, *125*

L

Laclau, E., 12, 71, *125*
Lamberton, D. M., 57, *125*
Lash, S., 73, 73(n2), 96, 97, *125*
Layder, D., 93, 114, *125*
Lee, C.C., 73(n2), *125*
Lent, J., 76, 76(n3), 78, 78(n5), *125*
Lerner, D., 28, 82, 83, *125*
Lummis, D., 41, *125*

M

Malik, M., 30, 31, 43, 53, *125*
Marglin, S., 108, *125*
Martin, G., 88, 89, *126*
Martín-Barbero, J., 91(n10), 93, 100, 101, 108, *126*

Martinussen, J., 99, *126*
Matta, F.R., 90, 91, 92, 93, *126*
Mattelart, A., 8, *126*
McAnany, E., 33, 34, *126*
Melucci, A., 100, *126*
Mills, M. A., 30(n5), 73, 85, 86, 96, *121*
Mohanty, C.T., 67, 100, *126*
Mosco, V., 14, 109, *125*
Mouffe, C., 12, 71, *125*

N

NAMEDIA, 8, 15, 16, 73, 77, 81, 83, 84, 108, *126*
Nandy, A., 19, *126*
Narula, U., 67, *126*
Nawaz, S., 35, *126*
Negri, A., 14, *124*

O

O'Sullivan-Ryan, J., 90(n8), *126*
Ogan, C., 36, 37, *126*
Oliveira, J. M., 33, 34, *126*

P

Pal, Y., 78, 79, 90, 108, *126*
Parthasarathi, G., 77, 84, 86, 89, *126*
Pavlic, B., 13, 41, 96, *126*
Pearce, B., 67, *126*
Pelton, J., 54, *126*
Pieterse, J.N., 107, *126*
Pool, I., 12, 16, 76, 90(n9), *127*
Poster, M., 46, *127*

R

Raboy, M., 117, *127*
Radhakrishnan, R., 120, *127*
Rahnema, M., 42, 45, 58(n3), 59, 91, 101(n2), 110, *127*
Randall, V., 70, *127*
Ranganath, H., 31, 53, *127*
Reeves, G., 92, *127*
Rist, G., 100, *127*
Robertson, R., 73, 74, 96, 98, *127*
Rodriguez, C., 10, 90, 103, *127*
Russell, A., 108, *127*

S

Sachs, W., 37, 64, *127*
Said, E., 9, 15, 27, 74, 88, 114, *127*
Saukko, P., 99, 120, *127*

Saunders, K., 10, *127*
Savio, R., 5, 84, 85, 85(*n*4), *127*
Schiller, H., 49, 96, 97, *127*
Schudson, M., 120, *127*
Sen, J., 101, *128*
Sen, M., 84, 101, *125*
Servaes, J., 10, *128*
Shah, H., 2, 4, 7, 10, 11, 14, 24, 36, 37, 41, 69, 75, 80, 108, 112, 116, 120, *123*, *128*
Shiva, V., 99, *128*
Shohat, E., 17, *128*
Sikkink, K., 108, *125*
Singham, A.W., 72, 72(*n*1), *128*
Slater, D., 9, 103, 104, *128*
Sosale, S., 99, 119, *128*
Spivak, G.C., 9, 30, 120, *128*
Spurr, D., 17, 41, 48, 49, 50(*n*1), 55, 56, 63, 64, 71, *128*
Sreberny-Mohammadi, 96, *128*
Stam, R., 17, *128*
Steeves, L., 10, *128*
Stevenson, R., 55, *128*

T

Tehranian, M., 1, 10, 11, 86, 87, 88, 89, 90, 101, 101(*n*1), *128*
Theobald, R., 70, *127*
Thomas, C., 97, 98, 99, *129*
Thussu, D., 8, 69, 76, 76(*n*3-4), 77, 96, 109, *122*, *129*
Tipps, D., 58(*n*3), *129*
Tomlinson, J., 96, 98, *129*

Traber, M., 14, 69, *130*
Tuchman, G., 36, 120, *129*
Tunstall, J., 36, 73(*n*2), *129*

U

UNESCO, 3, 4, 5, 7, 8, 14, 15, 16, 21, 22, 23, 24, 27, 28, 39, 40, 43, 45, 46, 50, 55, 56, 57, 59, 63, 69, 70, 81, 86, 89, 96, 115, 119, *129*
United Nations, 26(*n*3), 27, 39, *129*
United Nations Development Programme, 109, 110, *129*
Urry, J., 73, 73(*n*2), 96, 97, *125*

V

Venturelli, S., 20, *129*
Verghese, G., 28, 43, 81(*n*6), *129*
Vilanilam, J., 76, 76(*n*3), 78, *125*
Vincent, R., 14, 69, *129*, *130*
Volosinov, V.N., 103, *130*

W

Walsham, G., 29, *122*
Waters, M., 73, 73(*n*2), *130*
White, H., 16, 17, *130*
Williamson, H.A., 100, *130*
World Bank, 109, 111, *130*
World Bank Group, 109, *130*

Y

Young, R., 9, 44, 89, *130*

Z

Zizek, S., 12, 16, *130*

SUBJECT INDEX

A

Aboriginal television
  Australia, 120
Acceptance
  difference, 108
Access, 92
Accountability
  development through
    communication, 47–67
  progress reports, 57–60
  project evaluations, 57–60
  Third World, 47–67
ACLF. *See* Audio Cassette Listening
  Forum Project (ACLF)
Address, inclusive, 39–42
Adult communicators, 63–65
AFP. *See* Agence France-Presse (AFP)
Africa
  communication needs, 41
  famine, 24
  financial resources, 65
  media access, 110
  National Remote Sensing Center, 34
  satellite technology, 64
  space programs, 65
Agence France-Presse (AFP), 76
Agronomy, 86
Air transport, 97
Alcoholism, 106
Algeria journalists, 76

Alternate media theory, 102
Alternative communications, 89–91
Alternative language, 92
Ambiguity, 11, 12, 71
AP. *See* Associated Press (AP)
Archaeological foci
Archives official meetings, 26–27
Arena of contestation, 17
Asia media access, 110
Aspirations of developing nations, 22
Associated Press (AP), 76
Audio Cassette Listening Forum Project
  (ACLF), 44, 58
Australia aboriginal television, 120
Autonomy, 87

B

Balanced communication order, 22
Balanced exchange, 39
Balanced flow, 39
Big Four, 76
Bloch, Maurice, 27
Bolivia
  ethnographic studies, 120
  participant observation, 120
Bonn Consequences for Development
  Policy, 35
Bourdieu's concept
  Third World countries, 15
Brazil development projects, 59
Broadcasting, 28

Buenos Aires Remote Sensing
  Applications and Satellite
  Communications, 35
Bulletin boards, 92

C

CANA. *See* Caribbean News Agency
  (CANA)
Canto Nuevo, 92
Capitalism
  global
    information technology and, 119
    rural development, 119
  western, 97–98
Caribbean media access, 110
Caribbean News Agency (CANA), 76
Cellular telephony, 115–116
Center
  constructing, 76–77
  reconfiguring, 73–74
Centering, 11
Change
  social, 39
    mapping discourse, 9–13
Chile music, 92
Civilization universalization, 74–75
Classes in literacy, 106
Classifications of social intelligibility, 9
Collective self-reliance, 75
Colonialism, 87
  discourse, 48
Communication
  African needs, 41
  alternative, 89–91
  conceptions, 47
  contextualized nature, 60
  counterimaginings, 70–71
  deferred society, 65–66
  democratic, 4
  dominant and dominated, 91
  and economics integration, 56
  formulation, 40
  global
    balanced flow, 39
    connections, 107–108
  GNP, 56
  indicators, 55
  international scholarship, 11

  issue, 4
  knowledge, 55
  mapping discourse, 9–13
  mass, 86
  media
    infrastructures and institutions,
      27–32
    View from India, 27
  and mystique, 65–66
  negotiating readings, 69–94
  non-Western culture, 89
  NWICO vs. WSIS, 118–119
  order
    balanced, 22
    new world, 118–119
  overseeing, 49–50
  perspectives, 10
  practices
    expressive cultural practices, 42–45
    participatory cooptation, 42–45
  project proposals, 23
  research development lessons, 55–57
  returning sociality, 85–86
  rural satellite technology, 54
  satellite, 28, 51–54, 97, 119
    and related technology, 33, 41
    vocabulary, 53
  social, 86, 108
  socioeconomic development, 55
  technologies, 108
  theory, 31
  third world accountability, 47–67
  understanding, 81–82
  visual forms, 52
  women participatory project, 58
Communication and development, 72
  conceptions, 47
  discursive strategies, 16
  genealogy of idea, 14
  global era, 11, 111
  history, 48
  international era, 111
  organization and management, 33–38
  projects and proposals, 21–46
  reconceptualization, 101
  rural satellite technology, 54

Subject Index

Communication Policy and Development
  progress and impact, 40
Communication Rights for the Information Society (CRIS), 3, 117–118
Communicators, 63–65
Communities
  fragmentation, 120
  participatory communications, 43–44, 45
  rural Kenya, 103
  virtual grassroots, 119
Community Media Methodology, 57
Computer, 115–116
Conscientization, 43
Consequences for Development Policy Bonn, 35
Constructing
  center, 76–77
Contestation, 71
  arena of, 17
  discursive, 11
Contextualized communication, 60
Cooptation
  participatory communication practices, 42–45
Counterimaginings
  world order and communication, 70–71
CRIS. *See* Communication Rights for the Information Society (CRIS)
Critical genealogy, 10
Cultural autonomy, 87
Cultural development, 39
Cultural traditions
  expressive, 43
  expressive communication practices, 42–45
Culture and globalization, 96–103

D

Dag Hammarskjöld Third World Journalists' Seminar, 76, 83–84
Data purchasing, 53
Debates
  NWICO, 8
Deconstruction
  surveillance, 49
  unity of fields, 11
Deferred promise development, 61
Deferred society
  communications and mystique, 65–66
Dehumanizing nature technology, 87–88
Democracy
  developmentalized, 95
  implications, 2
  NWICO, 3–9
Democratic communication, 4
DepthNews, 76
Developing nations
  desires and aspirations, 22
Development, 13, 23, 109–110. *See also* Communication and development
  Brazil projects, 59
  communication
    conceptions, 47
    indicators, 55
    projects guiding, 62–63
  cultural, 39
  deferred promise, 61
  defined, 44
  and democracy, 3–9
  globalization, 1
  Guatemala projects, 59
  issue, 4
  mapping discourse, 9–13
  new, 115–120
  NWICO, 3–9
  overseeing, 49–50
  power and visibility, 48
  projects, 48, 59
  rural
    global capitalism and information technology, 119
    telephones, 29
  social, 39
  socioeconomic, 55
  strategies, 4
  symbolized, 64
  third world accountability, 47–67
  through communication, 47–67
Developmentalized democracy, 95

Subject Index

Development lessons
  communications research, 55–57
Dialogue, 88
Difference acceptance, 108
Digital media, 119
Discourse
  colonialism, 48
  description, 17
  exclusionary practice, 13
  mapping, 9–13
  modernization negotiations, 13
  tropological analysis, 17
Discourses
  Foucault's analyses of, 9
Discursive contestation, 11
Discursive space, 14–17
  policy making efforts, 14
Discursive strategy
  communication and development, 16
  generalizing, 38
  MacBride Report, 39
Disparities
  technological, 99
Documentary video
  India's women social movement, 105–106
Dominant and dominated
  world order and communication, 91
Dominant order
  challenges, 114–115
Drishti Media Group, 107

E

East African women, 24
East Asia
  NIC, 98
Economics
  integration, 56
  planning
    knowledge, 57
    resource control, 57
  transparent, 51–54
Edutainment, 24
Egypt
  traditional media roles, 30, 32
Electronic learning society, 89
Emancipation
  defined, 107–108
Enter-educate, 24
Enumeration, 21, 23
  methods, 25
  strategies, 25–26
Ethiopia
  INPE's management style, 34
Ethiopian and Somalian famines
  women, 24
Ethnographic studies, 120
Europe media access, 110
Evaluation
  media technologies, 12
Event, 37
Exchange balance, 39
Exclusionary practice discourse, 13
Expressive cultural practice, 42–45
Expressive cultural traditions, 43

F

Famine, 24
Fiberoptic cable, 97
Field of visibility, 59
Field reality, 59
Film, 103
Financial resources, 65
Financial transactions, 97
First and Third Worlds differences, 42
Floating signifier, 12
Foucault's analyses of discourses, 9
Foucault's Eurocentric archaeological foci, 9
Foucault's Francocentric archaeological foci, 9
Fragmentation, 120
Framing, 48
Free trade, 97–98
Freire, Paolo, 43

G

Genealogy
  communication and development, 14
  critical, 10
Generalization strategies
  Third World nations, 38
Generalized rhetoric, 39–42
Generalizing discursive strategy, 38
Global capitalism
  information technology and rural development, 119

Subject Index

Global communication
  balanced flow, 39
Global connections
  communication, 107–108
Global era
  communication and development, 11, 111
Global information and communication
  balanced flow, 39
Globalization, 71–72, 95–112
  and culture, 96–103
  development, 1
  uneven, 99
GNP. *See* Gross National Product (GNP)
Graffiti, 92
Grassroots virtual communities, 119
Gross National Product (GNP), 56
Guatemala development projects, 59

H

Habitus, 15, 24
Hidden overseers, 60–61
History
  communication and development, 48
  loss, 100
Humanistic journalism, 120
Human progress, 54

I

ICANN. *See* Internet Corporation for Assigned Names and Numbers (ICANN)
ICT. *See* Information and communication technologies (ICT)
Idea
  communication and development, 14
  gap, 33
  genealogy, 14
  satellite communications and related technology, 33
Identity loss, 100
Ideology defined, 114
Image enhancement, 52
Impact
  Communication Policy and Development, 40
Inclusive address, 39–42
India
  agronomy, 86
  communication media, 27
  journalists, 76
  traditional media roles, 30
  women's social movement documentary video, 103, 105–106
Indigenous communication theory, 31
Inequality, 98
Information, 57
  balanced flow, 39
  expanding concepts, 83–84
  global, 39
  international market, 36
  new international, 4
  policy, 4
  technology rural development, 119
  third systems of, 85
  Third World media professionals, 36
  UNESCO intergovernmental deliberations, 4
Information and communication technologies (ICT), 108, 115
Infrastructures and institutions
  communication media, 27–32
INPE's management style, 34
Institutions
  communication media, 24, 27–32
Intelligibility classifications, 9
Intelsat, 54
International communication
  scholarship, 11
International era
  communication and development, 111
International information market
  Third World media professionals, 36
International Telecommunication Union (ITU), 3, 115–116, 119
Internet, 99, 115–116
Internet Age, 118
Internet Corporation for Assigned Names and Numbers (ICANN), 119
Interpretation invisibility, 61
Intersections, 22
Invisibility, 48, 60–61
  interpretation, 61

methods, 62
themes, 62
ITU. *See* International Telecommunication Union (ITU)

J

Japan
   telephone vs. letters, 89
Japan-China-Southeast Asia triangle, 78
Journalism
   humanistic, 120
   restructuration, 79–80
Journalists
   Algeria, 76
   Dag Hammarskjöld Third World Journalists' Seminar., 76, 83–84
   India, 76
   Mexico, 76
   Pakistan, 76
   Peru, 76

K

Kamiriithu Community Educational and Cultural Center (KCECC), 104–105
Kenya
   literacy program, 104–105
   play, 104–105
   radio, 63
   rural community, 103
Knowledge
   communications, 55
   economic planning, 57

L

Language
   shared alternative, 92
Latin America media access, 110
Learning society
   electronic, 89
Letters *vs.* telephones, 89
Literacy, 103
   classes, 106
   program, 104–105
Local media, 103–107

M

MacBride, Sean, 14
MacBride Commission studies, 55
MacBride Report, 4
   discursive strategy, 39
MacBride Round Tables, 15, 69
Macrospeak, 21, 23, 37–38
   methods, 38
   themes, 38
Management styles, 34
Mapping
   discourse, 9–13
   unity of fields, 11
Mass, 103
Mass communication, 86
Mass media
   modern, 102
   nonmodern mobilization, 43
Master signifier, 12
M'Bow, Amatar, 23
Media
   access, 110
   communication infrastructures and institutions, 27–32
   View from India, 27
   digital, 119
   institutions, 24, 27–32
   local, 103–107
   mass, 102
   modernizing rural areas, 59
   new, 115–120
   ownership, 109–110
   traditional roles
      Egypt, 30, 32
      India, 30
Media professionals
   Third World, 36
Media technologies
   society evaluation, 12
Media theory, 102
Mediation replacing media, 108
Mexico journalists, 76
Mobilization
   nonmodern mass media, 43
Modernization, 13, 88, 98
   and development, 23
      symbolized, 64
   discourse negotiations, 13
   mass communication, 86, 102

Subject Index 141

peripheral, 13
rural areas media, 59
vernacular, 13
Western, 99
Movements
  social, 103–107
Music
  Chile, 92
Mystique, 61
  deferred society, 65–66

N

NAM. *See* Nonaligned Movement (NAM)
NAMEDIA, 108
  goals, 83–84
NAMEDIA Conference, 8
  publications, 15
  reports, 77
Naming, 100
NANAP. *See* Nonaligned News Pool (NANAP)
National Remote Sensing Center
  African countries, 34
National system
  seamless integration, 57
Nations
  developing desires and aspirations, 22
  parent adult communicators, 63–65
Natural dehumanizing technology, 87–88
Negotiating readings
  world order and communication, 69–94
Negotiations, 71
  with modernization discourse, 13
NEMS. *See* News exchange mechanisms (NEMS)
New International Economic Order (NIEO), 12, 77–78
New international information and communication order, 4
New international information policy
  UNESCO intergovernmental deliberations, 4
Newly Industrialized Countries (NIC), 98
New media, new developments, and old questions, 115–120
News bulletin boards, 92
News exchange mechanisms (NEMS), 76
New world communication order
  NWICO *vs.* WSIS, 118–119
New World Information and Communication Order (NWICO), 4
  debates, 8, 75, 83, 113–114
  peak, 14
  reading development, 14–17
  development, 3
  democracy, 3–9
  economic subjugation history, 7
  new world communication order, 118–119
  policy, 6
  political history, 7
  proponents, 69
  vs. WSIS, 118–119
NGO. *See* Nongovernmental Organization (NGO)
Ngugi wa Mirii, 104–105
Ngugi wa Thiong'O, 104–105
NIC. *See* Newly Industrialized Countries (NIC)
NIEO. *See* New International Economic Order (NIEO)
Nonaligned Movement (NAM), 72
Nonaligned News Pool (NANAP), 76
Nonalignment movement, 4
Nongovernmental Organization (NGO), 12, 98
Nonmodern mass media mobilization, 43
North America media access, 110
NWICO. *See* New World Information and Communication Order (NWICO)

O

Observation participant, 120
OECD. *See* Organization for Economic Cooperation and Development (OECD)
Official meetings
  rituals and archives, 26–27
Opinion expanding concepts, 83–84

Orality, 102
Orbital space allocations, 29
Organization for Economic Cooperation and Development (OECD), 110
Overcrowding satellite stationing space, 29
Overseeing communications, 49–50
Overseeing development, 49–50
Overseers hidden, 60–61
Ownership, 109–110

P

Pacific media access, 110
Pakistan
    journalists, 76
    press, 80
Pan-African News Agency (PANA), 76
Panoptic view, 48
    power, 49
Parent nations
    adult communicators, 63–65
Participant observation
    Bolivia, 120
Participation, 103
Participatory communication project
    women, 58
Participatory community communications, 43–44, 45
Participatory cooptation communication practices, 42–45
Peripheral modernity, 13
Periphery, 75
    reconfiguring, 73–74
Peru journalists, 76
Policy making efforts, 14
Political and economic subjugation history NWICO, 7
Postdevelopment, 109–110
    communities, 103–107
    and globalization, 95–112
Power
    panoptic view, 49
    types, 49
    and visibility in development, 48
Practice
    communication practices, 42–45
    exclusionary discourse, 13
    expressive cultural, 42–45

Pradesh, Andhra, 105–106
PrepComs (Preparatory Committee meetings), 118
Press
    Pakistan, 80
Print, 103
Progress
    accountability reports, 57–60
    Communication Policy and Development, 40
    human, 54
    social, 39
Projects
    development, 48, 59
        communication, 62–63
    evaluation accountability, 57–60
Promise deferred development, 61
Public opinion, 83–84
Purchasing data, 53

R

Radio
    Kenya, 63
    satellite, 115–116
Rasa Siddhanta, 31
Reading development, 14–17
Reality field, 59
Recalling resolutions, 27
Relations of ruling establishment, 24
Remote Sensing Applications and Satellite Communications, 35
Remote-sensing centers, 52
Replacing media, 108
Report accountability, 57–60
Research communications, 55–57
Resource control, 57
Restructuration, 71, 73–74
    journalistic interventions, 79–80
    method, 75
    themes, 75
Retheorizing, 72, 81–82
    methods, 82
    themes, 82
Returning sociality communication, 85–86
Rhetoric, 39–42
Rituals of official meetings, 26–27
Rural areas modernizing media, 59

Subject Index

Rural communication development, 54
Rural community, 103
Rural development, 119
   telephones, 29

S

SACI/EXERN project
   1967-1974, 33
Satellite-based networks
   Third World countries, 79
Satellite communication, 51–54, 97, 119
   and broadcasting, 28
   and related technology, 41
      idea gap, 33
      vocabulary, 53
Satellite radio, 115–116
Satellite stationing space overcrowding, 29
Satellite technology
   Africa, 64
   rural communication development, 54
Schematic exaggerations, 17
Self-reliance
   collective, 75
   women, 44
Sen, Mrinal, 84
Services WTO, 7
Shared alternative language, 92
Social change, 39
   mapping discourse, 9–13
Social communication, 86
Social development, 39
Social intelligibility
   classifications of, 9
Sociality, 86, 89
   returning communication, 85–86
Social movement, 103–107
   women's
      documentary video, 105–106
      India, 103
Social nature communication, 108
Social progress, 39
Society deferred communications, 65–66
Society evaluation of media technologies, 12
Socioeconomic development, 55
Somalian famines, 24
Song and Drama Division, 31

South Asia media access, 110
Space
   discursive, 14–17
   programs, 65
Strategies
   development, 4
   discursive
      communication and development, 16
      generalizing, 38
      MacBride Report, 39
   enumeration, 25–26
   Third World nations, 38
Structuration, 73–74
Subjugation history
   political and economic NWICO, 7
Sub-Saharan Africa media access, 110
Surveillance, 48, 49–50
   deconstruct, 49
   methods and activities, 50
   themes, 50
Symbolized modernization, 64
Systems of information, 85

T

TAI. *See* Technology achievement index (TAI)
Tanzania
   ACLF, 44
   USAID, 58
Techno-economic valorization, 87
Technology
   dehumanizing nature, 87–88
   disparities, 99
   media society evaluation, 12
Technology achievement index (TAI), 109
Telephones, 30
   cellular, 115–116
   vs. letters, 89
   rural development, 29
Television, 120
Theater, 103
   traveling, 92
Themes
   invisibility, 62
   restructuration, 75
   retheorizing, 82
Third World, 6, 22

accountability, 47–67
  Bourdieu's concept, 15
  development through communication, 47–67
  difference between First World, 42
  international information market, 36
  media professionals, 36
  satellite-based networks, 79
  strategies of generalization, 38
  training programs, 81
TNC. *See* Transnational Corporations (TNC)
Trade, 97–98
Traditional expressive culture, 43
Traditional media roles
  Egypt, 30, 32
  India, 30
Training programs, 81
Transnational Corporations (TNC), 96, 98
Transparent economies, 51–54
Transportation, 97
Traveling theater, 92
Tropological analysis
  discourse, 17

U

UNDP. *See* United Nations Development Programme (UNDP)
UNESCO, 3, 15, 96
  intergovernmental deliberations
    new international information policy, 4
  new international information and communication order, 4
United Nations Development Programme (UNDP), 109
United Nations motto, 2
United Press International (UPI), 76
Unity of fields
  deconstruction, 11
  mapping, 11
Universal civilization, 74–75
UPI. *See* United Press International (UPI)
US Agency for International Development (USAID)
  Tanzania, 58

V

Valorization techno-economic, 87
Vernacular modernity, 13
Video
  women's social movement, 105–106
View from India, 27
Virtual communities
  grassroots, 119
Visibility
  in development, 48
  field of, 59
Visual communication forms, 52
Vocabulary satellite communication, 53

W

WB. *See* World Bank (WB)
Western capitalism, 97–98
Western modernity, 99
WGIG. *See* Working Group on Internet Governance (WGIG)
When Women Unite: The Story of an Uprising, 105
Women
  East Africa, 24
  Ethiopian and Somalian famines, 24
  India, 103
    documentary video, 105–106
  participatory communication project, 58
  self-reliance, 44
  social movement, 105–106
Working Group on Internet Governance (WGIG), 118
World Bank (WB), 109
World communication order
  counterimaginings, 70–71
  dominant and dominated, 91
  negotiating readings, 69–94
  NWICO vs. WSIS, 118–119
World Summit on the Information Society (WSIS), 3, 115–117
  vs. NWICO, 118–119
World Trade Organization (WTO)
  services, 7
WSIS. *See* World Summit on the Information Society (WSIS)
WTO. *See* World Trade Organization (WTO)

Printed in the United States
92791LV00002B/34-81/A